PLANT PROCESSING OF NATURAL GAS

Issued by
PETROLEUM EXTENSION SERVICE
The University of Texas at Austin
Austin, Texas

in cooperation with
TEXAS EDUCATION AGENCY
Department of Occupational Education and Technology
Austin, Texas

and

COMMITTEE ON VOCATIONAL TRAINING
American Petroleum Institute
Division of Production
Dallas, Texas

and

GAS PROCESSORS ASSOCIATION
Tulsa, Oklahoma

1974

Published 1974. Eleventh Impression 1995
Printed in the United States of America

Brand names, company names, trademarks, or other identifying symbols appearing in illustrations or text are used for educational purposes only and do not constitute an endorsement by the author or publisher.

The University of Texas at Austin is an equal opportunity institution. No state tax funds were used to print or mail this publication.

Catalog No. 3.11010
ISBN 0-88698-115-8

Table of Contents

Foreword

This is the first edition of *Plant Processing of Natural Gas* and is intended to be used in conjunction with *Field Handling of Natural Gas,* which is also published by Petroleum Extension Service. The information in this manual describes the latest procedures, practices, and equipment used in the plant processing of natural gas, and it was collected by a special group headed by H. J. "Hank" Haas, chairman of the Advisory Committee for the School of Gas Technology. This school, as well as a number of others, is operated by Petroleum Extension Service at Kilgore College and is jointly sponsored by API Division of Production and GPA.

Grateful acknowledgment is made to the following people for their work in compiling and reviewing the material in this manual:

Ken H. Bacon—Warren Petroleum Company
Clem A. Barrere, Jr.—Continental Oil Company
Lloyd E. Childers—Amoco Production Company
Eugene Damron—Exxon Company, U.S.A.
J. Gordon Drost—Cities Service Company
Freeman L. Morgan, Jr.—Consultant
Carl E. Springer—Continental Oil Company
Ervin W. Squires—Exxon Company, U.S.A.
Willard Sutton—Houston Natural Gas Corporation

C. F. Kruse
Petroleum Extension Service
The University of Texas at Austin

Austin, Texas
November, 1974

Preface

The purpose of this manual is to present the fundamental concepts and workable practices for the processing of natural gas. The manual has been prepared primarily to aid the plant employee who deals regularly with the problems of processing and treating gas.

The special review committee appointed to validate this edition has been mindful of this purpose and has endeavored to include positive, useful information, while minimizing highly technical material that would not contribute to the productive capability of the plant employee.

The manual does not cover all materials, practices, procedures, or equipment used in the processing and treating of natural gas. Also this book should not be interpreted as indicating a preference by the committee for one specific manner of operation. The reader must be guided by his own experience in dealing with variances in local conditions and requirements.

It has been necessary and desirable to include photographs and drawings describing equipment that may be easily identified as to manufacturer. The committee did not intend by these inclusions to infer in any way a preference or practice that is limited to equipment manufactured by any specific company.

Henry J. Haas, Chairman
API-GPA Validation Committee

Gas Processing Plant Equipment

Chapter I

Fundamentals

Whenever you see a pipe, a steam line, a tank, a pump, a compressor, a tower, an instrument, or even a filled sample container in a gas plant, it almost certainly contains a fluid. What is a fluid? Anything that can flow is a fluid. Gases (or vapors) and liquids are different forms of a fluid. Fluids have no definite shape. A fluid assumes the shape of the vessel into which it is placed.

Solving operating problems, increasing operating efficiency, and avoiding operating hazards nearly always require an understanding of how fluids behave. Anytime a change in operation is to be made, four questions need to be asked and answered:

- What is the fluid?
- What are the properties of the fluid?
- What are we attempting to do with the fluid?
- What are the present conditions of the fluid?

FLUIDS

Whenever a problem occurs in any part of the process, it is likely to be related to the laws and properties of fluids. Every fluid has properties that distinguish it—make it different—from other fluids in a given situation. Some properties by which fluids may differ from each other are—

- temperature
- pressure
- gravity
- miscibility
- solubility
- phase
- vapor pressure
- boiling point
- freezing point
- hydrate point
- heat capacity
- flammability
- composition

Before reviewing the various properties of fluids and how operating conditions influence fluid behavior, an understanding of fluid composition is necessary. Everything is composed of atoms. Most atoms can combine with other atoms. These combined groups of atoms are called molecules. Each fluid contains many, many molecules. How a fluid behaves depends on how its molecules behave. Picture a glass jar filled with marbles, as in figure 1.1. Each marble represents a molecule. All the molecules together form the fluid that is contained in the jar. Actually, molecules are much too small to be seen even with a microscope.

The important fact to remember is that each fluid has many small parts—its molecules. If the molecules can be made to do what is wanted, then the fluid will do what is wanted. Process engineers use molecules in the design calculations for a gas processing plant.

In the gas processing industry, temperature is expressed in degrees Fahrenheit. This is generally written as degrees F. Each fluid has a property called *critical temperature.* If a fluid is at a temperature greater than its critical temperature, then the *phase* of the fluid—that is,

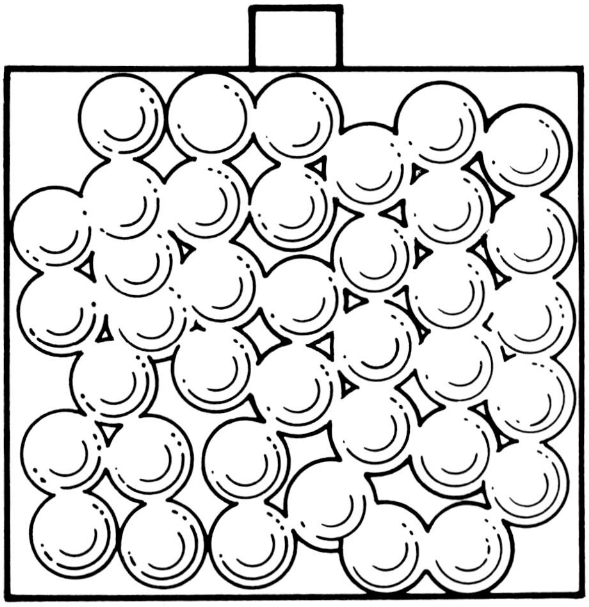

Figure 1.1. Fluid molecules may be compared to marbles.

whether the fluid is a gas or a liquid—will depend on its pressure.

Sometimes in the estimation of fluid behavior, a temperature scale called degrees Rankine (R) has to be used. Degrees Rankine is an absolute temperature scale and is equal to 460 plus the degrees F. For example, a fluid at 100 F is also at 460 plus 100 F, or 560 R. This, plus some other temperature scales, is illustrated by figure 1.2. Absolute temperatures are used in engineering calculations concerned with gas behavior, such as compressor cylinder capacity and compressor discharge temperatures.

Pressure is the weight or force in pounds applied on one square inch of area, and therefore it is expressed in pounds per square inch (psi). For example, if a 1-in. square block of wood were placed on the floor and a 200-lb man stood on the block of wood, then the pressure applied to the floor would be 200 psi.

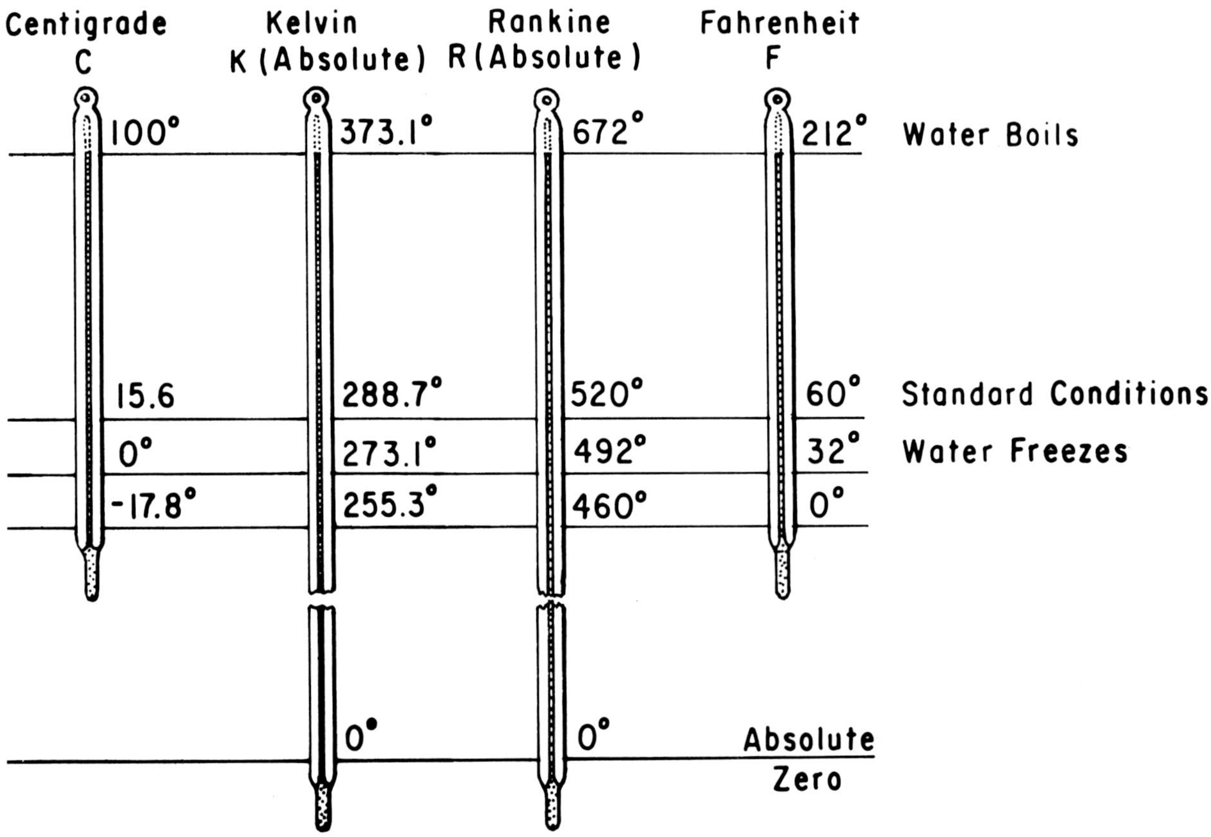

Figure 1.2. Comparison of Centigrade, Fahrenheit, and Absolute Temperatures

Pressure in a tank is usually indicated on a gauge. The pressure shown by the gauge is called *gauge pressure* and is abbreviated psig—pounds per square inch gauge. Although psig is a useful measure, it is not the true total pressure. The pressure gauge on the tank shows the difference between the pressure in the tank and the pressure of the atmosphere. The true or absolute pressure in the tank is the gauge pressure plus the atmospheric pressure. The absolute pressure is called psia, pounds per square inch absolute. At sea level, the atmospheric pressure is 14.7 psi. So, the absolute pressure in the tank is the gauge pressure (psig) plus 14.7. Psia equals psig plus 14.7. Fluid vapor pressures are correlated using psia.

Each fluid has a property called *critical pressure.* Critical pressure is important in gas processing because a liquid mixture cannot be separated if the pressure is above the critical pressure of either the mixture or the separated fluids.

Another property of a fluid is *gravity.* Gravity is the heaviness of the fluid. The heaviness of a hydrocarbon liquid, such as propane, gasoline, or oil, is customarily measured in degrees of API gravity. The lighter the liquid, the higher its API gravity. Specific gravity relates the density—that is, a fluid's weight per unit volume—of one fluid to the density of another. Liquids are usually related to water, and gases are usually related to air.

When two liquids do not mix, and the lighter liquid floats on top of the heavier liquid, they are said to be *immiscible*—that is, they do not mix. If oil and water are put into the same tank, they will separate, if given enough time, so that the oil floats on the water. Almost all of the water can be drained from the tank, leaving only the oil. Therefore, oil and water are immiscible. When two liquids mix without separating they are said to be *miscible.* If butane and gasoline are put into the same tank, they will not separate even though butane has a lower specific gravity; in other words, butane is lighter than gasoline. Because butane and gasoline are miscible, it is not possible to separate the two liquids by draining the heavier, higher-specific-gravity gasoline from the tank.

All hydrocarbons in a gas processing plant are miscible; they will mix with each other. On the other hand, hydrocarbons are immiscible—they will not mix—with fluids such as water and glycol and monoethanolamine, or MEA. (MEA, a liquid used in the processing of gases that contain sulfur, will be discussed later). A problem with mixtures of hydrocarbons and water, hydrocarbons and glycol, and hydrocarbons and MEA is *solubility.* To understand solubility, consider the case of a gasoline and water mixture. Even though the two fluids are immiscible, some water will dissolve in the gasoline, and some gasoline will dissolve in the water. In other words, to a small extent some water is soluble in gasoline, and some gasoline is soluble in water. In this case, the solubility is not much—in fact, less than 1/2 lb of water will dissolve in each 1,000 lb of gasoline. However, at low temperatures, even this small amount of water could result in freezing problems. Similarly, glycol and hydrocarbons are soluble. The solubility of glycol in hydrocarbons and hydrocarbons in glycol can create problems. One problem is that considerable quantities of glycol can be lost due to hydrocarbon contamination.

The *phase* of a fluid is important. A fluid can be in a gas, liquid, or solid phase. Gas processing is concerned with two phases of fluids: gas and liquid. Incidentally, gases are sometimes called *vapors;* a vapor is the same thing as a gas. A gas, or vapor, consists of millions of molecules rapidly moving in all directions. These molecules collide with each other and with everything else that attempts to contain or hold them, such as the walls of a vessel.

Gas pressure in a vessel is caused by the impacting force of the molecules colliding with the walls of the vessel. Anything that is done to either decrease the number of collisions or distribute the number of collisions over a larger area will reduce the gas pressure. Anything that is done to either increase the number of collisions or reduce the area of collisions will increase the gas pressure. What does this mean? It means that gas pressure in a closed vessel can be controlled two ways: by changing either the temperature or the volume of the gas. If the gas is heated to raise its temperature, its molecules move faster. They strike each other and the vessel walls more often, and the pressure in the vessel increases. If a gas is cooled, its molecules move more slowly and the pressure decreases. If a gas is expanded into a larger vessel, the distance among gas molecules increases. The molecules do not collide with each other as often, and they have a larger vessel wall area with which to collide. This means that the pressure of the gas decreases. If the gas is compressed, its molecules are forced to occupy a smaller volume. This means that the impact of the gas molecules is concentrated on a smaller area. Therefore, the gas pressure increases.

If there is no liquid present, then all gases have the same pressure if they are in vessels of the same sizes and at the same temperatures. Of course, if gas were in a vessel open to the atomosphere, and if an attempt were made to increase the pressure, some of the gas would escape to equalize with the pressure of the surroundings. However, it might be possible to increase the gas pressure within the vessel so fast that the gas could not escape fast enough through the available opening to equalize the pressure inside and outside of the vessel.

The molecules of a liquid are held close together by a natural attraction between the molecules. This attraction between molecules tends to keep each molecule confined to a small space. However, within this space a molecule is usually in violent motion. In a partially filled vessel, the violent motion of the molecules at the surface of the liquid causes many of these molecules to escape from the liquid. The molecules that escape form a vapor in the space above the liquid surface. As a result, some of the gas molecules strike the surface of the liquid; those that strike the liquid surface tend to stick there. Thus, molecules may be escaping the liquid and returning to the liquid from the gas at the same time, as illustrated by figure 1.3. When the number of molecules leaving the liquid equals the number of molecules entering the liquid, the gas and liquid are said to be in *equilibrium* with each other. This does not mean that the gas volume is equal to the liquid volume. It only means that the number of molecules moving from gas to liquid and from liquid to gas is equal. When the liquid phase and the vapor phase are at equilibrium, the pressure of the vapor is said to be the vapor pressure of the liquid. When the temperature of a liquid in a closed vessel is increased, the pressure of the vapor increases until the vapor and liquid are at equilibrium with each other. Therefore, when the temperature of a liquid in a closed vessel is increased, its vapor pressure is increased. (See fig. 1.4.)

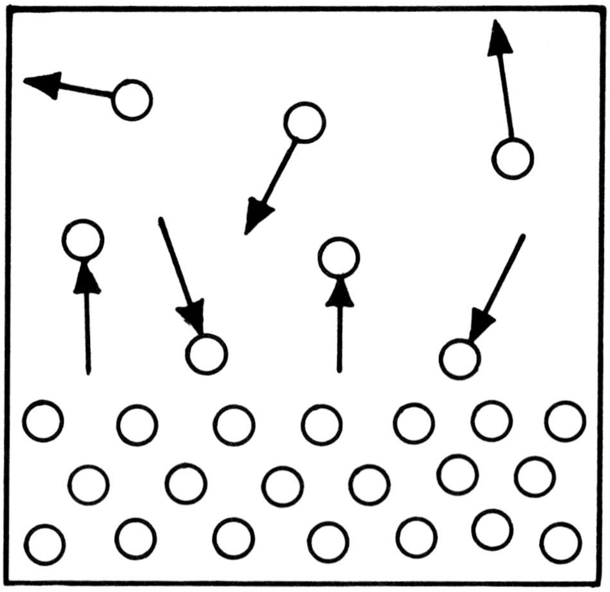

Figure 1.3. Molecules escape and return to the liquid phase in closed vessel.

Pressure is controlled in distillation and fractionation towers. Only in unusual situations is the pressure varied. Therefore, for these types of operations, the relation of vapor pressure and temperature needs to be thought of in a different way. A liquid boils when its vapor pressure becomes equal to the pressure within the vessel. If the vessel pressure is fixed, then for a liquid to boil its temperature must be increased. If a liquid is boiling and it is desirable to stop the boiling action, then the temperature of the liquid must be lowered.

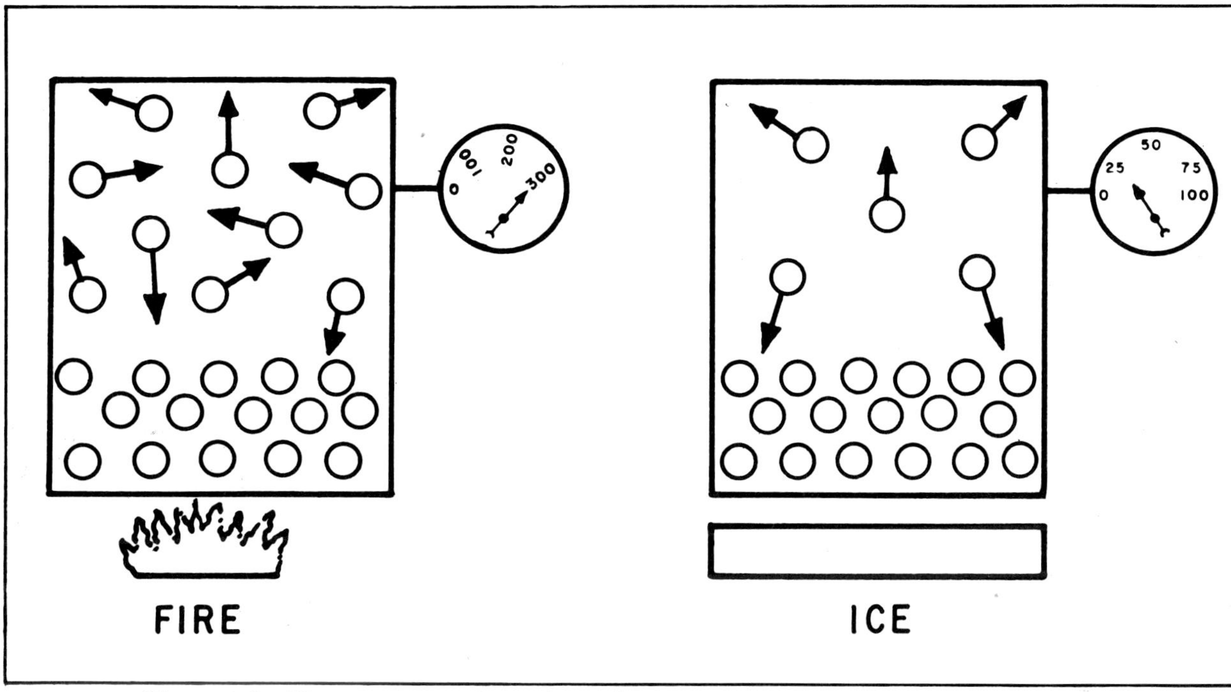

Figure 1.4. Vapor pressure depends on the temperature of a liquid in a closed vessel.

For each liquid, at each pressure, there is a temperature at which the liquid will boil. This is called the *boiling point,* or *boiling temperature.* If the pressure is increased, the temperature must be increased for the liquid to continue boiling. Of course, another way to say this is that at each temperature there is a pressure—that is, the vapor pressure of the liquid—at which the liquid will boil. Figure 1.5 is a graph of vapor pressure versus boiling temperature for a liquid.

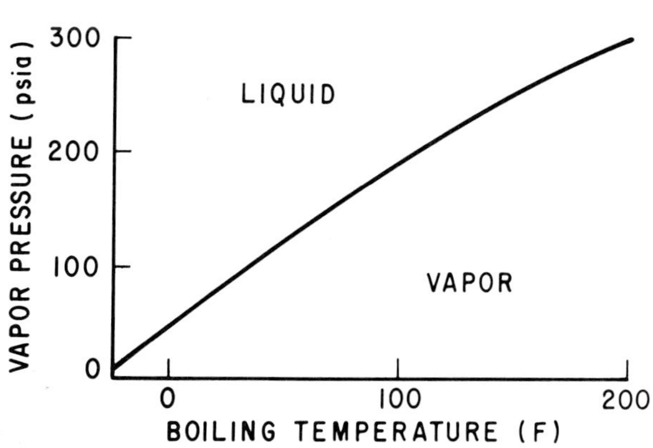

Figure 1.5. Vapor Pressure Compared to Boiling Temperature of a Liquid in Closed Vessel

When more molecules are escaping from the liquid than are returning, the liquid is vaporizing (becoming a vapor). When more molecules are entering the liquid from the vapor than are escaping the liquid, the vapor is condensing. A change in the phase of a fluid, from either vapor to liquid or liquid to vapor, is only a change in the distance between molecules. The molecules of a gas are farther apart than the molecules of a liquid. A change in phase does not change the molecular structure or the chemical nature of the fluid.

When a liquid is heated and is free to expand, it will expand in all directions. For this reason, liquid expansion is called volume expansion. Many operators have bottled a sample of gasoline when it was cold and then had the cork pop off the bottle when the gasoline warmed to room temperature. The effects of volume expansion can be much more dangerous when handling light liquids such as propane or butane. If a pressure vessel is filled completely, the volume expansion of the material could rupture the vessel. Since the content is liquid under pressure, immediate vaporization would occur, causing an explosion. This is why the vapor must be permitted to vent freely.

The Gas Processors Association (GPA) has specified a 20 percent outage when filling pressure containers at temperatures above 0 F. This means that in taking a sample at least 20 percent of the sample container should be left as vapor space for the volume expansion that will occur when the sample container is brought into a warm room. The Interstate Commerce Commission (ICC) has set specifications on how full a tank car can be loaded with a liquid petroleum, or LP, gas such as butane. These specifications make an allowance for volume expansion.

In gas processing operations, problems created by the freezing point of pure hydrocarbon vapors and liquids are not generally of much concern because these temperatures are very low. However, if *free water* is present in the vapors and liquids, freezing problems often occur even at moderately low temperatures—that is, those above 32 F. (Free water is water that could be drained from a vessel and thereby separated from the gas or liquid.) Even though the freezing point of the gas or liquid might be very low, the presence of free water can cause *hydrates* to form. Hydrates are an icy mixture of hydrocarbons and water that may form at temperatures as high as 80 F, depending on the pressure. Hydrates can plug lines, valves, heat exchanger tubes, and even vessels. Hydrate plugs can usually be eliminated by injecting methanol or glycol, increasing the temperature, or lowering the pressure. However, the best method of handling hydrate problems is to prevent their formation by careful operation.

The hydrocarbons of primary concern in gas processing operations are listed in figure 1.6. Also listed are some of their more important physical properties. This figure shows that the hydrocarbons have been listed with the lightest at the top and the heaviest at the bottom of the table. The abbreviated symbol for each hydrocarbon is listed immediately following the name. These are not the correct chemical designations but are the abbreviated symbols commonly used in gas processing. The next column lists the specific gravity of the hydrocarbon when in a liquid phase at 60 F. Because the specific gravity of a liquid changes with tem-

Hydrocarbon	*Symbol*	*Liquid Specific Gravity @ 60 F*	*Gas Specific Gravity*	*Boiling Point @ 14.7 psia F*	*Vapor Pressure @ 100 F psia*	*Critical Temp. F*	*Critical Pressure psia*
Methane	C_1	0.30	0.55	-259	5,000	-117	668
Ethane	C_2	0.36	1.04	-127	800	90	708
Propane	C_3	0.51	1.52	-44	190	206	616
Iso-butane	iC_4	0.56	2.01	11	72	275	529
Normal Butane	nC_4	0.58	2.01	31	52	306	551
Isopentane	iC_5	0.62	2.49	82	20	369	490
Normal Pentane	nC_5	0.63	2.49	97	16	386	489
Hexane	C_6	0.66	2.97	156	5.0	454	437
Heptane	C_7	0.69	3.46	209	1.6	513	397
Octane	C_8	0.71	3.94	258	0.5	564	361
Nonane	C_9	0.72	4.43	303	0.2	611	332
Decane	C_{10}	0.73	4.91	345	0.1	652	304

Figure 1.6. Physical Properties of Hydrocarbons Involved in Gas Processing

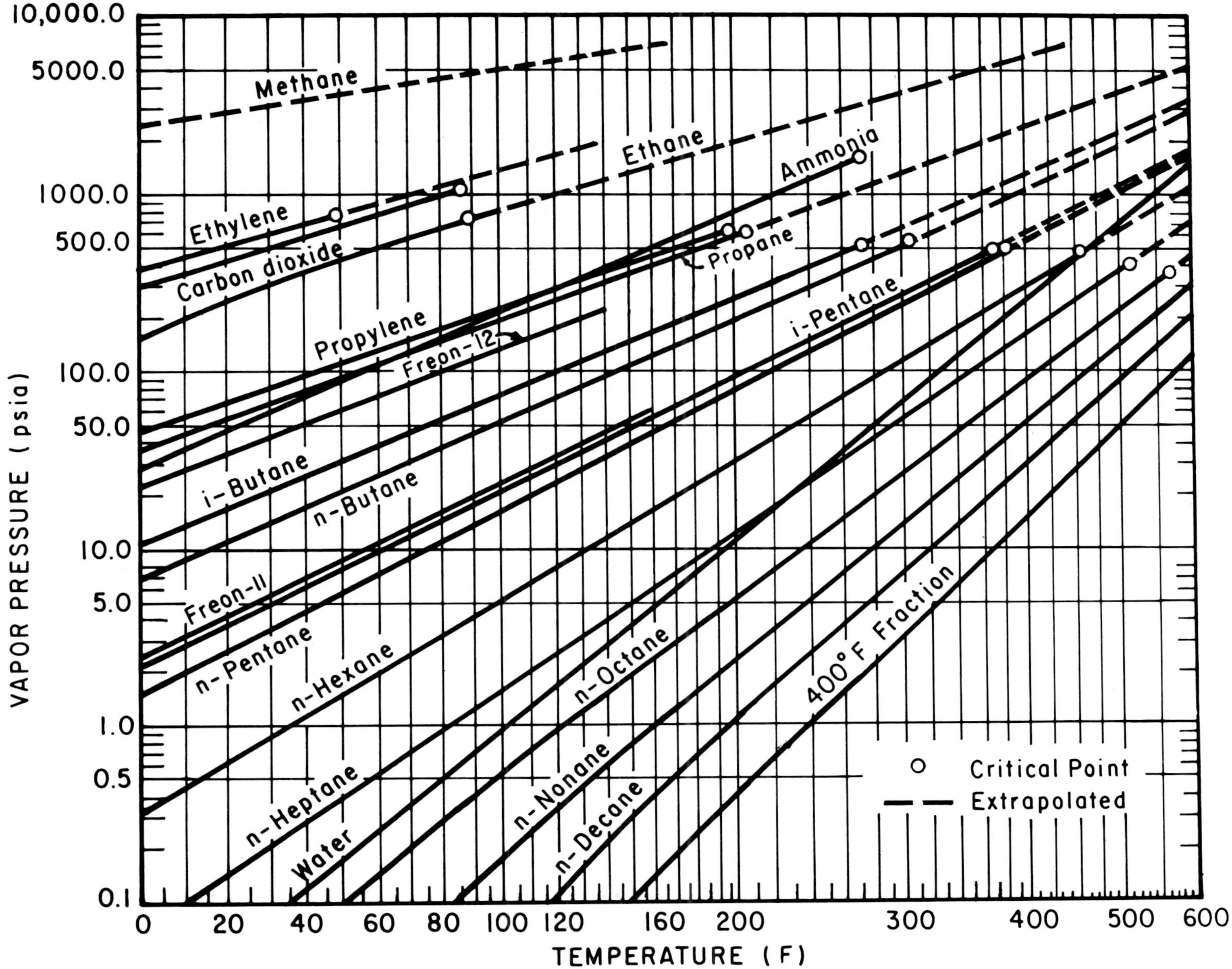

Figure 1.7. Vapor Pressures of Various Hydrocarbons

perature, due to the fact that liquids are lighter at higher temperatures, then anytime a specific gravity of a liquid is stated, the temperature of the liquid must be specified. The next column is the specific gravity of the hydrocarbon when it is in a gas phase. This is the weight of a specified volume of the gas compared to the weight of the same volume of air. Examination of this column shows that only methane, or a gas mixture with a lot of methane, is lighter than air since the specific gravity of air is 1.00. This means that if a vessel with mostly methane inside is vented, the vented vapor will rise. However, if a tank containing propane, or any of the heavier hydrocarbon gases, is vented, the vented vapor will settle to the ground and can create an unsafe condition. The next two columns of figure 1.6 list the boiling point at 14.7 psia (atmospheric pressure) and the vapor pressure at 100 F. Remember, the boiling point of a fluid depends on its pressure.

Figure 1.7 is a chart of vapor pressure versus boiling temperature for a number of hydrocarbons and other fluids. For example, if a propane surge tank is at 100 F, then, as shown on the graph, the vapor pressure of the propane is 190 psia. Figure 1.8 lists some of the physical properties of other fluids that are often used in gas processing operations.

In a mixture of gases, *mole percent* is often used to express the quantity of each gas present in the mixture. Mole percent is the ratio of the number of *moles* of one gas to the total number of moles; this number is multiplied by 100 so that the result will be on a percentage basis. For a mixture of gases, mole percent and volume percent are the same. However, mole percent is not equal to volume percent for a liquid mixture. A mole is the quantity of a substance the unit weight of which is numerically equal to the substance's molecular weight. For example, 1 pound mole of methane, CH_4, is 16 lb since the molecular weight of methane is 16. The unit weight could be expressed in grams, tons, or whatever unit is desired. Table 1 shows the mole percent of each component in a mixture of four gases using pound units. Note that the number of moles of each gas is determined by dividing the actual weight of the gas by its molecular weight since a mole is numerically equal to the component's molecular weight.

Other Compounds	*Liquid Specific Gravity @ 60 F*	*Boiling Point @ 14.7 psia F*	*Vapor Pressure @ 100 F psia*	*Freezing Point @ 14.7 psia F*	*Fire Point @ 14.7 psia F*
Nitrogen	0.81	-320	-	-	-
Carbon Dioxide	0.83	-109	1,250	-	-
Hydrogen Sulfide	0.79	-79	394	-	-
Water	1.00	212	0.9	32	-
Monoethanolamine					
(MEA) 100%	1.02	339	-	51	200
15%	1.01	213	-	20	-
Triethylene					
Glycol (TEG)					
100%	1.13	546	-	19	330
Ethylene Glycol					
100%	1.11	387	-	8	245
70%	1.09	240	-	-80	-

Figure 1.8. Physical Properties of Other Compounds

Table 1. Calculating Mole Percent

Component	*Symbol*	*Molecular Weight*	*Weight of Component*	*Calculated Number of Moles*	*Mole Percent*
Methane	C_1	16	60 lb	$\frac{60}{16} = 3.750$ moles	$\frac{3.750}{4.816} \times 100 = 77.87\%$
Ethane	C_2	30	20 lb	$\frac{20}{30} = 0.667$ mole	$\frac{0.667}{4.816} \times 100 = 13.85\%$
Propane	C_3	44	10 lb	$\frac{10}{44} = 0.227$ mole	$\frac{0.227}{4.816} \times 100 = 4.71\%$
Butane	C_4	58	10 lb	$\frac{10}{58} = 0.172$ mole	$\frac{0.172}{4.816} \times 100 = 3.57\%$
	Totals		100 lb	4.816 moles	100.00%

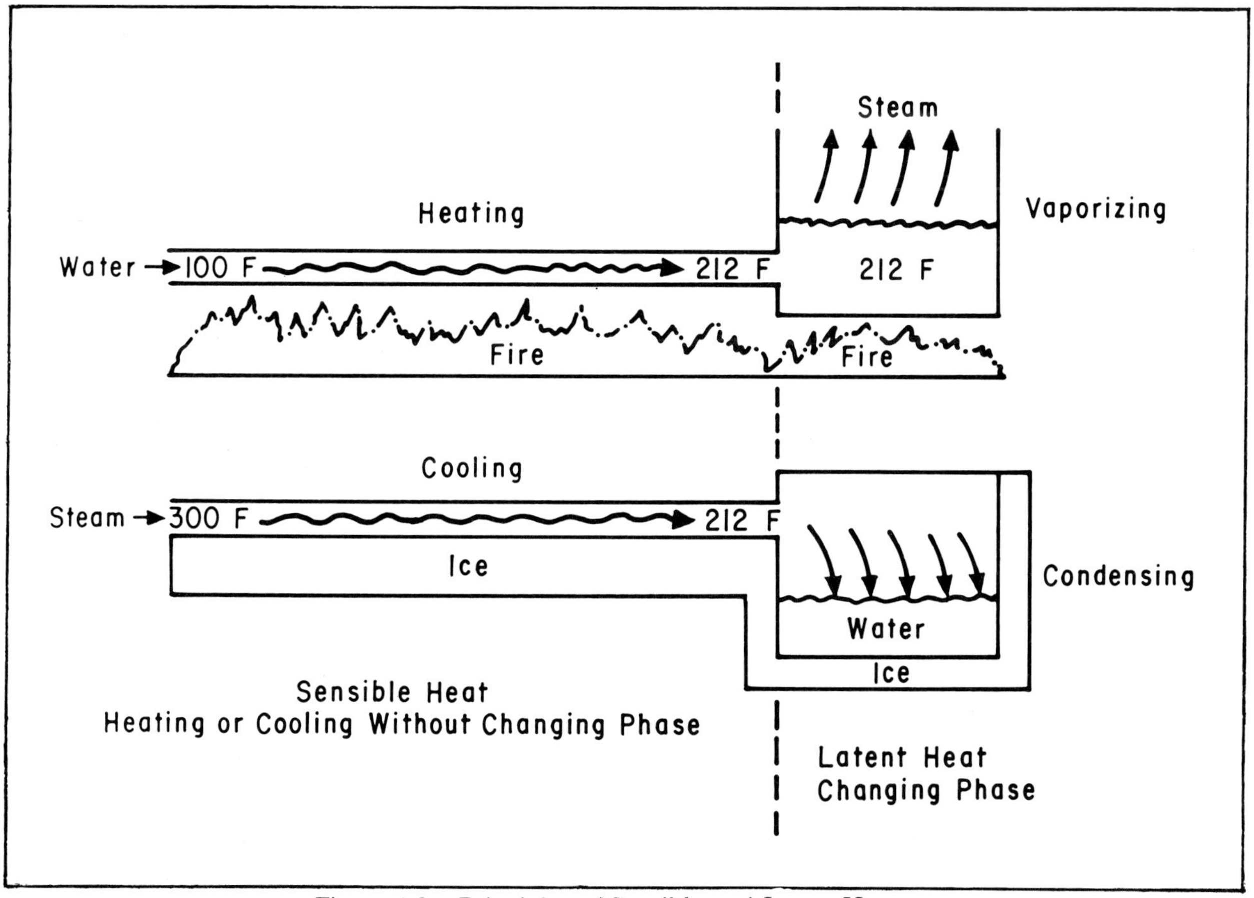

Figure 1.9. Principles of Sensible and Latent Heat

HEAT

There are three states of matter: solid, liquid, and gas. These three phases of matter are not permanent but depend upon the amount of heat energy contained by a given amount of substance. Thus, heat energy determines the physical state of matter of a given substance. If enough heat is added to an ice cube, which is a solid, it melts and changes to water, a liquid. If more heat is added to the water, it boils and changes into steam, which is a gas. Of course, heat does not always have to be added; heat may also be taken away, in which case the process of phase change is reversed. Water changes from steam to liquid to ice as the temperature is lowered.

The unit for measuring heat is called a British thermal unit (Btu). It is hard to imagine a Btu. The unit for measuring weight is the pound. A pound may be visualized because an object weighing a pound can be seen and handled; its weight can be felt. It is not possible to hold or see a Btu, but it is easy to understand what a Btu is. It is defined as the amount of heat that will increase the temperature of 1 lb of water by 1 F.

The heat that changes the temperature of a fluid without changing the phase of the fluid is called *sensible heat.* If heat is added to a fluid without changing its phase from liquid to gas, then the temperature of the fluid increases. Likewise, if heat is taken from a fluid without changing it from gas to liquid, then the temperature of the fluid decreases. The sensible heat capacity of a fluid is a measure of its ability to take on or let go of heat without changing its phase.

The heat that changes the phase of a fluid is called *latent heat.* If heat is added to a boiling liquid so that more vapor will be made, then the heat added is latent heat. If heat is taken from a condensing vapor so that more liquid will be made, then latent heat is being removed. Sensible heat and latent heat are illustrated in figure 1.9.

How much difference is there between sensible and latent heat? Both sensible heat and latent heat values depend on the fluid, its phase, its temperature, and its pressure. However, it generally takes about 250 Btu more heat to make 1 lb of vapor from a hydrocarbon liquid than it takes to raise the temperature of 1 lb of the hydrocarbon liquid 1 F. If vapor is being condensed, this ratio is still about the same. The difference for water is even greater. It takes about 1,000 Btu more heat to make 1 lb of steam from water than it takes to raise 1 lb of water by 1 F.

How is heat transferred? When a hot object is touched, the transfer of heat from the object can be felt because of the temperature difference between the hand and the object. An object feels cold to the touch when heat is transferred from the hand to the object. Heat energy is always transferred between solids, liquids, or gases if there is a temperature difference between them.

If a fluid at 200 F needs to be cooled to 120 F, and cooling water is available at 85 F, can the water be used for cooling? Surely, because there is a temperature difference. However, the amount of heat in the fluid depends upon both its temperature and the amount available—that is, the quantity of fluid available. The temperature difference means some cooling will occur, but there must be a sufficient quantity of water to cool the fluid quantity and still have some available temperature difference between the water and the fluid.

In summary, there must be both a temperature difference and sufficient quantity of fluid to heat or cool another substance. A fluid hotter than the heating medium can never be obtained; conversely, a fluid can never be cooled lower than the temperature of the cooling medium. The quantity of the heating or cooling medium must be large enough so that there is always a temperature difference to act as the driving force for heat transfer in the desired direction.

COMBUSTION

Combustion is the act or process of burning. Combustion is helpful in fired heaters, boilers, gas turbines, and internal combustion engines. Combustion is harmful when it occurs as an unwanted fire. The three essential requirements for combustion are—

1. a supply of oxygen;
2. a combustible fuel; and
3. heat energy, or ignition.

In both a fired heater and boiler, the main variable for combustion control is the amount of excess air supplied to the burners. This is because the two major heat losses are incomplete combustion and dry-gas loss. These are controlled by the relative amounts of fuel and air mixed for combustion.

The only way to find out exactly how efficiently a fuel is burning is to analyze the off gases (flue gas) from combustion and determine the concentration of the different products formed by combustion. Incomplete combustion means that the fuel is not completely burned. This loss of good fuel can be controlled by increasing the amount of excess air. Incomplete combustion is indicated by a flue-gas analysis that shows an oxygen content of 2.5 percent or less. A flue-gas content of 2.5 percent oxygen is equivalent to about 15 percent excess air. If there is too much excess air, great quantities of heat will be carried up the stack as the air passes through the furnace. This type of heat loss is called dry-gas, or stack, loss. If analysis of the flue gas reveals an oxygen content greater than 5 percent, which is equivalent to 30 percent excess air, this indicates that the stack loss was too great and the amount of excess air should be decreased. Too much excess air means increased fuel consumption for the same amount of useful heat.

Since combustion requires three essential elements, then a fire can be extinguished by—

1. cutting off the supply of fuel;
2. smothering the fire by taking away the source of oxygen; or
3. cooling the fire by removing the heat energy that keeps it going.

Not all vapor and air mixtures are flammable. Every gas has a specific flammable range within which it will ignite. This flammable range indicates the proportion of fuel to air that will ignite. At atmospheric pressure, the flammable range for gas processing hydrocarbons is about 1 percent to 15 percent—that is, mixtures between 1 percent hydrocarbon and 99 percent air and 15 percent hydrocarbon and 85 percent air. Below the lower limit of 1 percent hydrocarbon, there is not enough fuel for combustion. Above the upper limit of 15 percent hydrocarbon, there is not enough oxygen, and the fire will be smothered. Hydrocarbons will burn when ignited if the hydrocarbon vapor and air concentration is within the flammable range.

Normally, increasing the pressure will widen the flammable range of hydrocarbon and air mixtures. If there is any chance that air has entered a closed vessel containing hydrocarbons and the mixture may be explosive, the operation will be made safer by decreasing the pressure within the vessel. During depressuring, an explosive mixture should not be allowed to contact a source of ignition such as a flare.

The *flash point* of a liquid is the temperature at which only enough vapors are produced to cause a momentary flash when ignited. The lowest temperature at which a liquid produces enough vapors to burn continuously is called the *fire point.* If a liquid is burning, the burning gases will continuously add heat to the liquid. This added heat keeps the liquid vaporizing rapidly enough to support continuous combustion; that is, it keeps the liquid at its fire point temperature.

APPLICATIONS

A study of pure fluids allows the following conclusions:

- Light liquids have low boiling point temperatures and high vapor pressures.
- Heavy liquids have high boiling point temperatures and low vapor pressures.

Therefore, it can be said that light liquids tend to vaporize easily and that heavy liquids do not tend to vaporize easily. *Volatility* is the term used to describe the ease of vaporization of a fluid. Generally, the ease of vaporization of one fluid as compared to the ease of vaporization of another fluid is the main interest of plant operations; that is, the operators want to know the relative volatilities of the fluids being processed. Almost all gas processing operations are concerned with mixtures of fluids rather than just pure fluids. These fluid mixtures are composed of individual hydrocarbons. Therefore, knowledge of pure hydrocarbons and their relative volatilities is essential to good process operation.

When the vapor of a mixture is in equilibrium with the liquid of a mixture, all of the components present in the liquid are also present in the vapor. To say it another way, all of the components present in the vapor are also present in the liquid. This equilibrium situation of all components being present in both phases is the reason that gas processing has so many different operations. When removing the desired products from the inlet gas, some of the undesired components are also removed. In getting rid of the undesired components, some of the desired products are lost.

A pure fluid has a specific boiling point at a given pressure. At the boiling point temperature, a pure liquid will continue to boil until all of it has been vaporized. On the other hand, a mixture of several fluids will have a range of boiling temperatures. If the temperature of a mixture of several fluids is increased until it starts to boil, which is said to be the *initial boiling point,* some of the mixture will be vaporized. Boiling of the mixture will then stop unless the temperature is again increased. All of the mixture will not be vaporized until the *final boiling point* temperature is reached.

In a mixture of hydrocarbon fluids, the light molecules are able to escape the liquid more rapidly than the heavy molecules. The amount of light and heavy components left in a mixture of liquids is partially controlled by the temperature during the boiling operation. Therefore, as can be seen in figure 1.10, the light molecules tend to concentrate in the vapor, and the heavy molecules tend to concentrate in the liquid. It is this difference in volatility that makes gas processing possible.

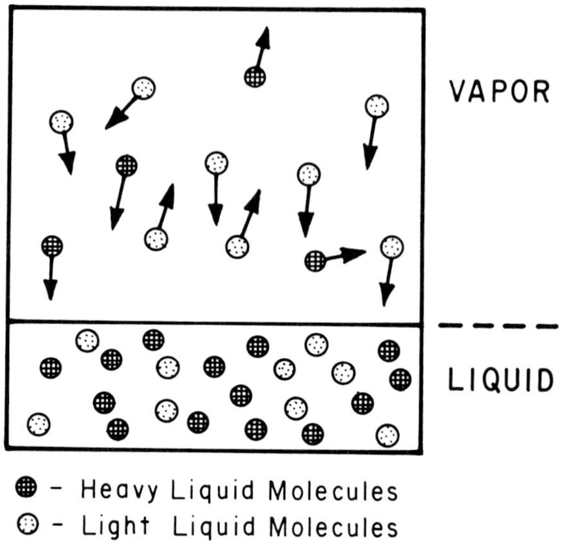

Figure 1.10. Light molecules vaporize, and heavy molecules concentrate in liquid.

Gas processing operations require that several items of equipment function as a unit. The connection of this equipment is normally illustrated by flow diagrams. Symbols are used on the flow diagrams to represent the various types of equipment. Figure 1.11 is a summary of the usual flow-diagram symbols. These symbols are used throughout this book.

Figure 1.12 is a flow diagram of a simple separation system. A fire is put under a liquid mixture, and part of the mixture is vaporized. Cooling water is injected into the vapor leaving the top of the first tank, and the vapor is condensed in another tank. The condensed liquid in the second tank is richer in the light component, and the liquid in the first tank is richer in the

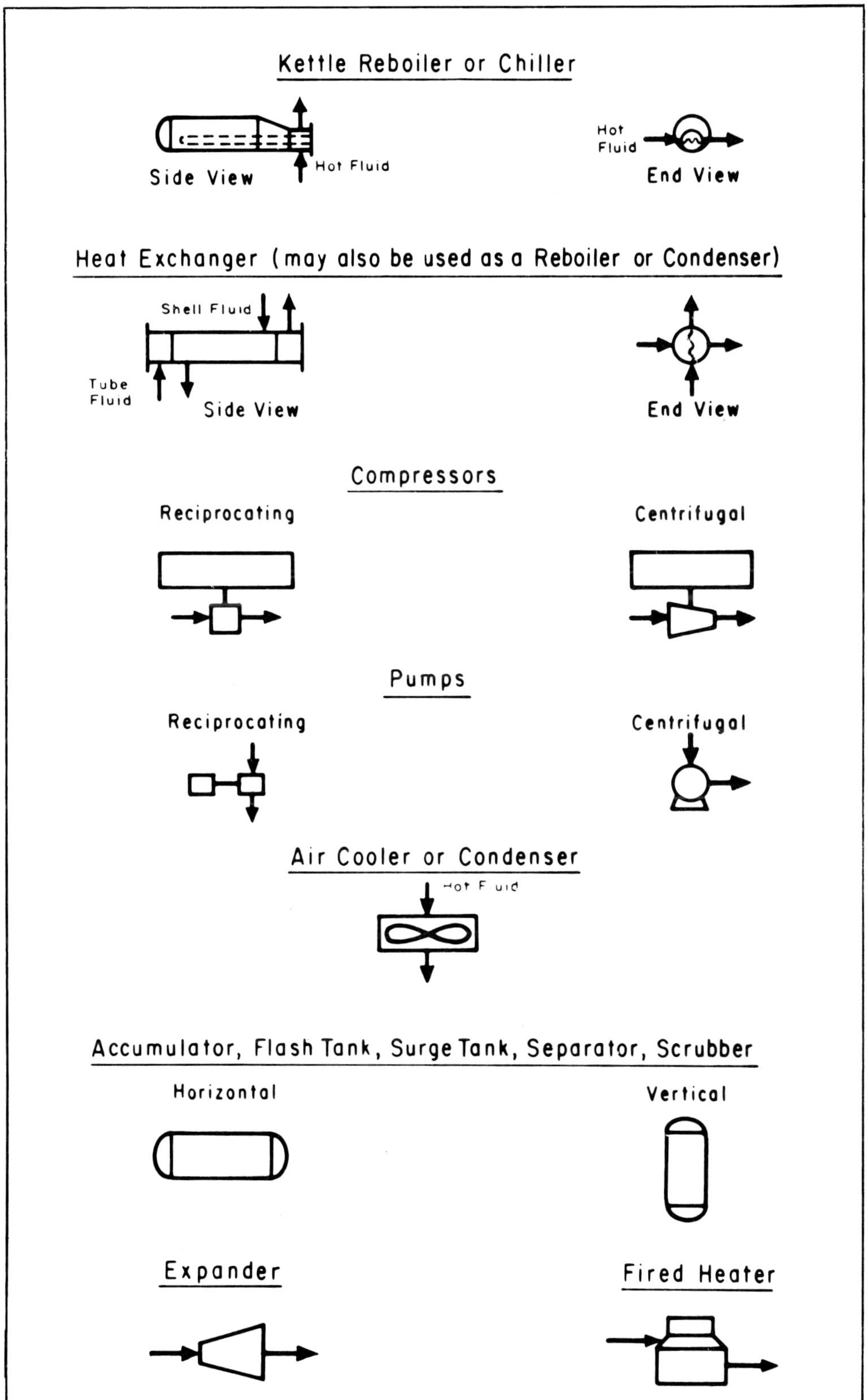

Figure 1.11. Flow-Diagram Symbols

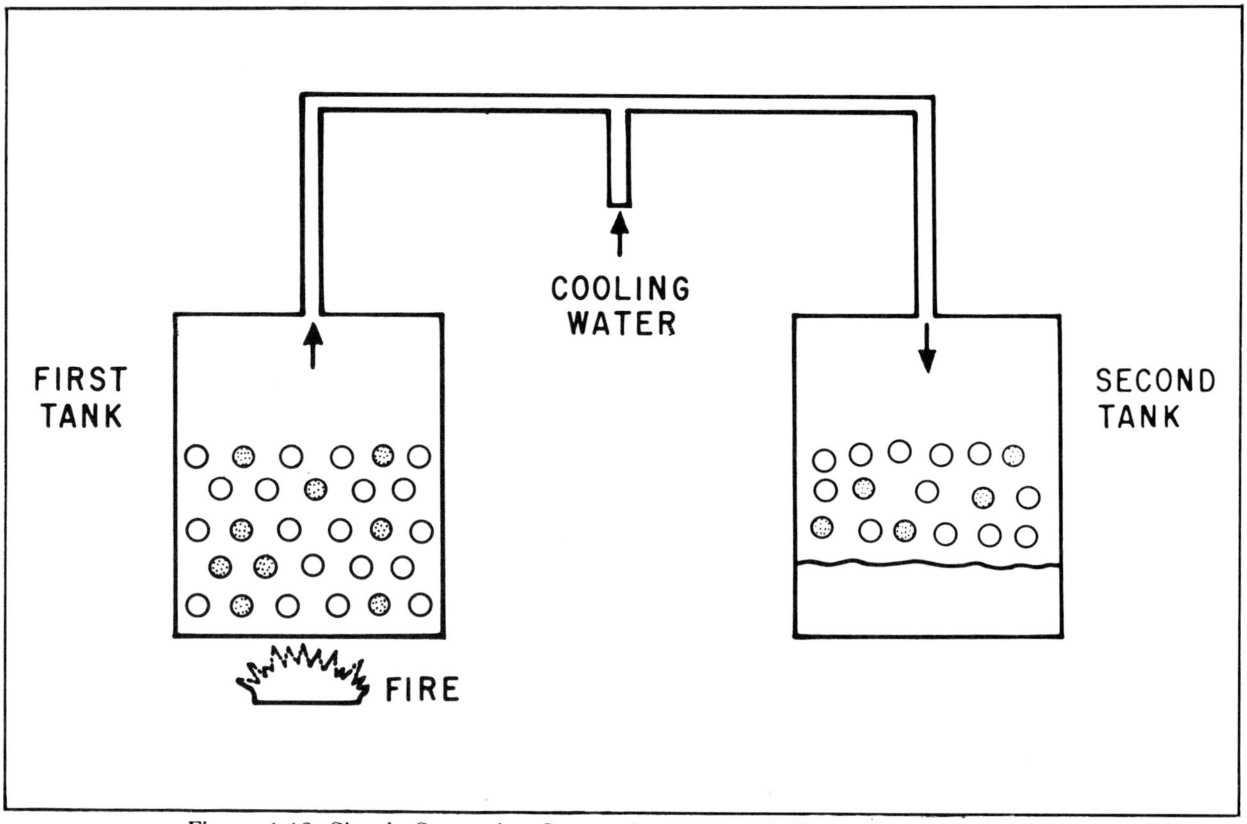

Figure 1.12. Simple Separation System for Light and Heavy Fluid Components

heavy component. This is a simplified system; however, there are a number of improvements that should be made. The open flame should be removed because it is a fire hazard. A reboiler, which uses steam to supply the needed heat, is a better source of heat, as shown in figure 1.13. Although a fired heater could be used, a shell-and-tube-type heat exchanger or a reboiler would be better in this instance. In a heat exchanger, the liquid mixture to be heated is in the shell, and the steam is in the tubes. A reboiler may have the liquid mixture in the tubes or in the shell and can employ either steam or hot oil for heating. There is another improvement that should be made because the cooling water is injected into the vapor and must be separated from the condensed light liquid. To avoid this, a shell-and-tube-type exchanger can again be used. Cooling water runs through the tubes, and the vapor is condensed in the shell. This piece of plant equipment is called a condenser, as illustrated by figure 1.14.

The separation system described previously is not very good because there is a lot of light component in the heavy liquid and a lot of heavy component in the light liquid. Several tanks in series, as shown in figure 1.15, will improve the separation. Now the separation system is reasonably efficient, but it is extremely complicated. It would also be very expensive to build.

Something is needed that will reduce the amount of equipment and yet allow efficient separation. The answer is a tower with trays. Trays used in towers are illustrated by figure 1.16. There are all kinds of trays: bubble cap, valve, sieve, perforated, and so forth. All of them do the same job—that is, they make possible good vapor and liquid contact. Good contact is necessary for good heat exchange and for good separation efficiency. Liquid flows across the tray and down a downspout, called a downcomer, to the next tray. Vapor flows up the tower and bubbles through the liquid on

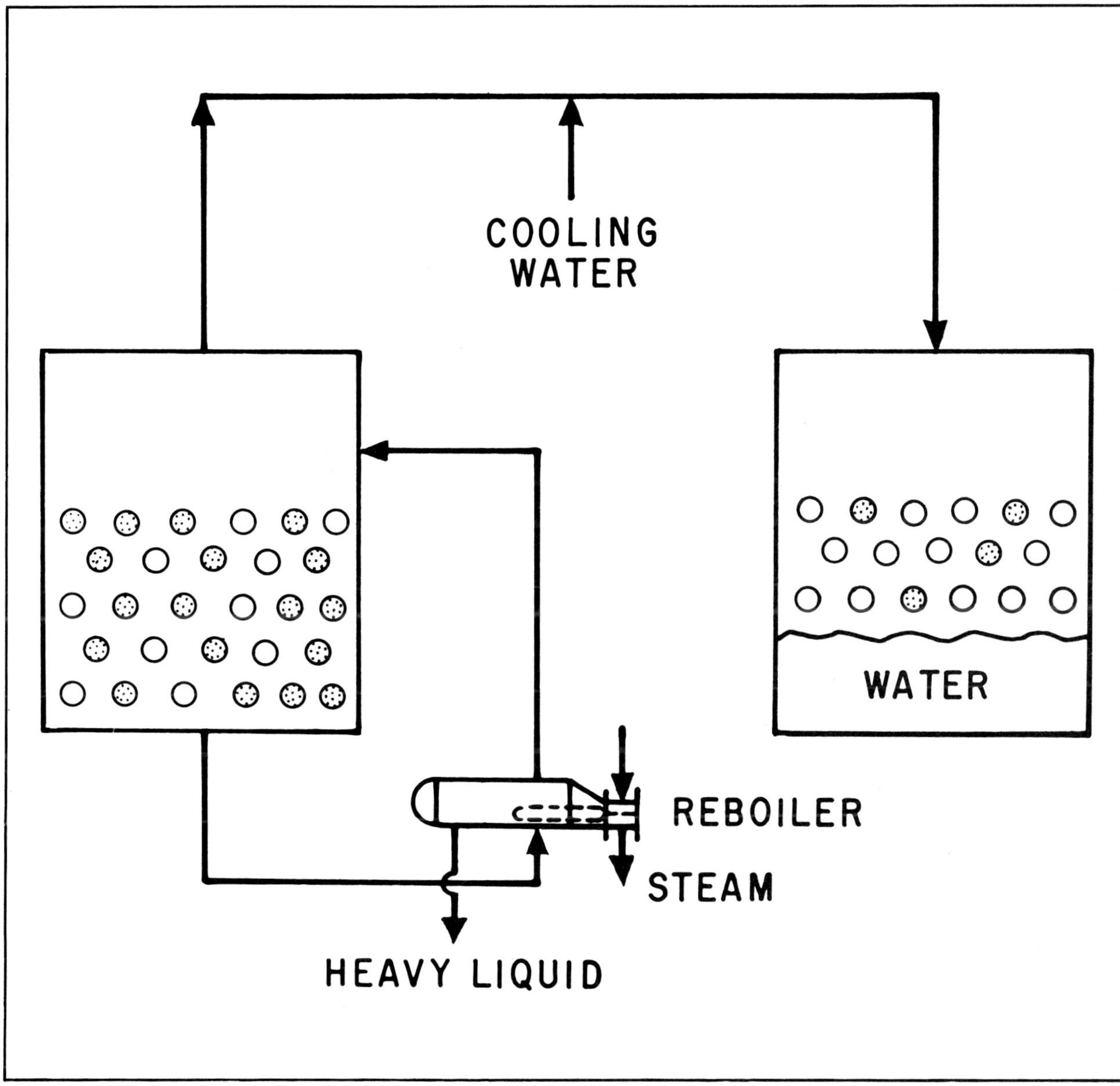

Figure 1.13. Simple Separation System with Reboiler

each tray. Trays are usually about 2 ft apart, and a number of trays can be put into a tower. Figure 1.17 shows a tower with the trays installed, with vapor rising in the tower through the liquid in the trays and the liquid falling in the tower and flowing across the trays until it reaches the bottom.

Figure 1.18 illustrates the new separation system using a tower with trays. The reboiler vaporizes the liquid. The reboiler temperature is set so that only a little of the light component is left in the heavy liquid. All of the liquid, other than the heavy-liquid product, that gets to the reboiler is vaporized. This vapor starts up the tower and contacts liquid coming down the tower. The hot vapor heats the cold liquid, and the cold liquid cools the hot vapor. In addition, the hot vapor causes some of the light component in the liquid to vaporize; that is, the hot vapor *strips* some of the light component in the

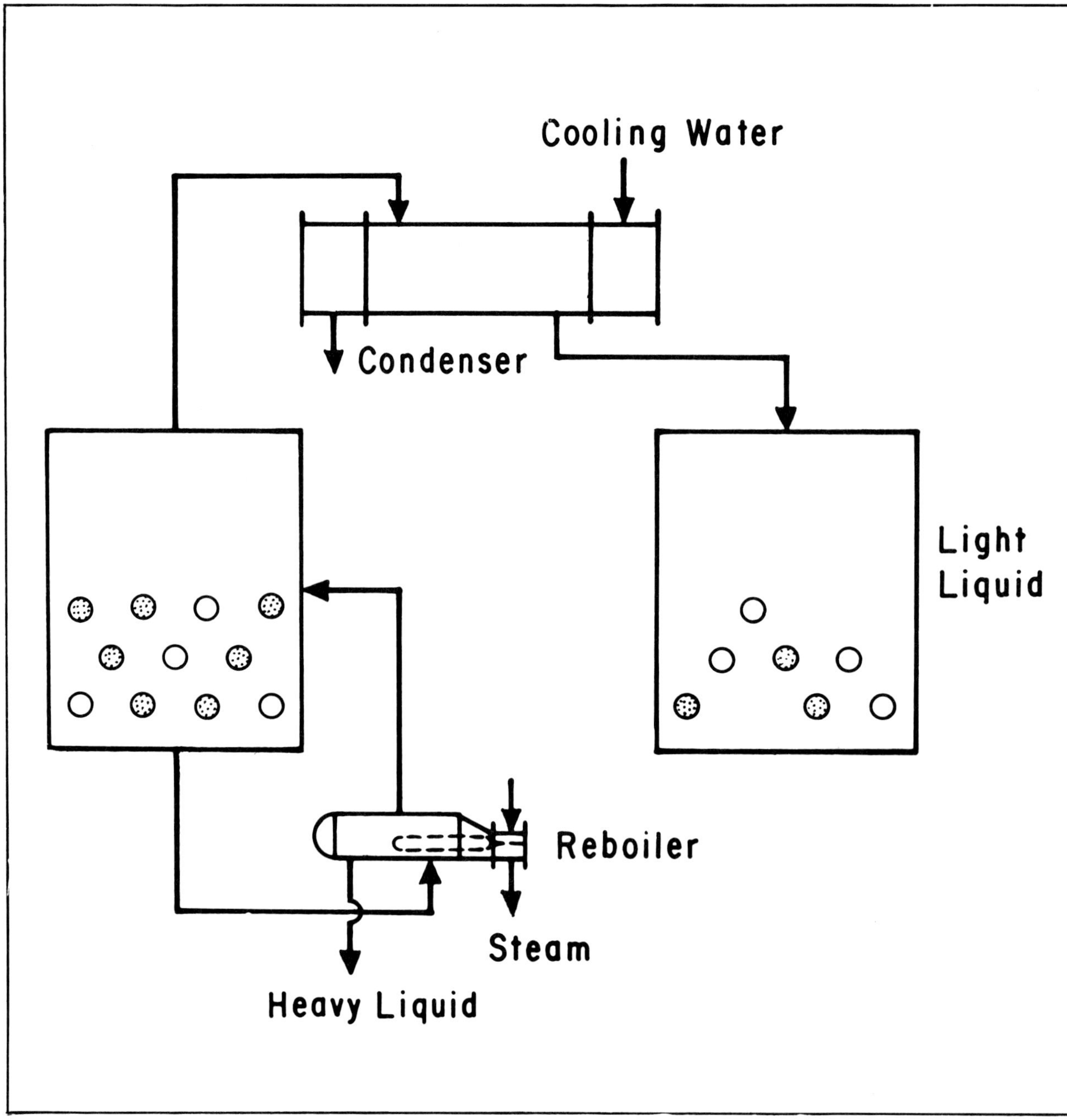

Figure 1.14. Simple Separation System with Reboiler and Condenser

liquid. For this reason, the bottom section of the tower between the feed and reboiler is called the stripping section.

The vapor flowing up the tower and past the point where the feed enters, however, still contains a lot of the heavy component. The heavy component must be kept out of the light-liquid product. So, part of the light-liquid product is pumped to the top tray of the tower. This stream is called *reflux.* The purpose of the reflux is to purify the light-liquid product. It does this by cooling the vapors that get to the tower's top section and thereby condensing the heavy component. The tower's top section above the feed is

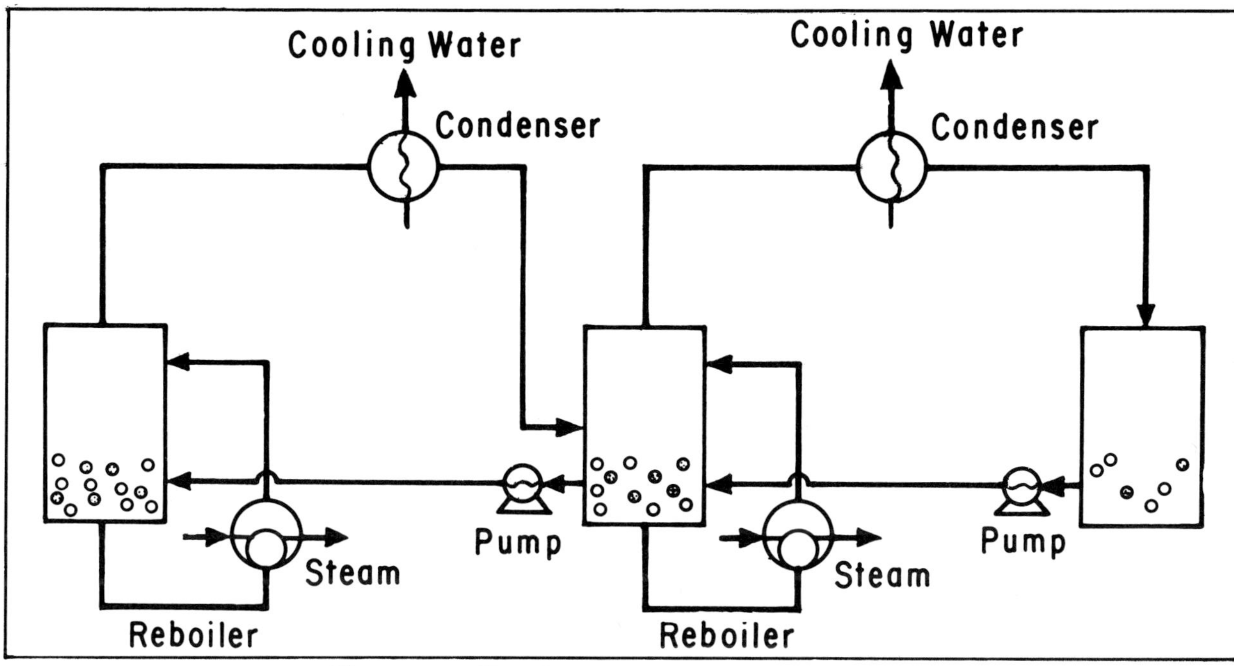

Figure 1.15. Flow Diagram of Series Separation System

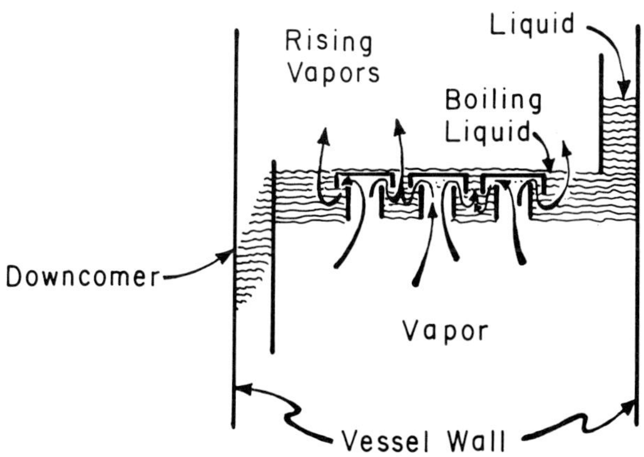

Figure 1.16. Bubble Cap Tray for Separation Tower

called the *rectifying,* or *rectification, section.* These are the elements of a good separation system. Every gas processing plant has facilities using these fundamental concepts.

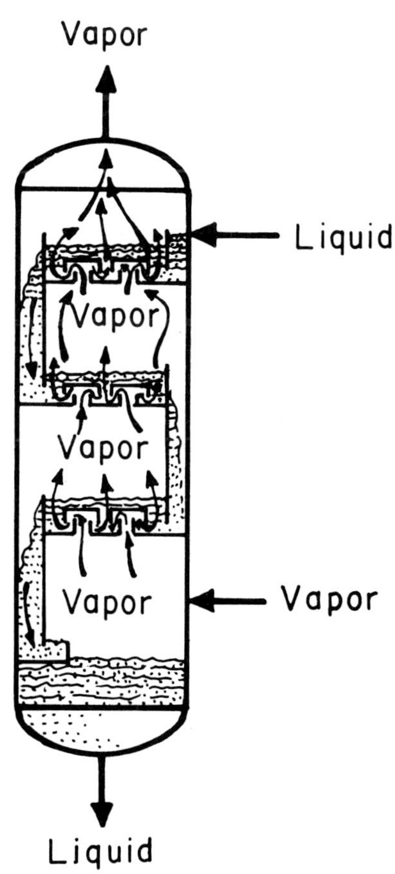

Figure 1.17. Bubble Trays in Tower

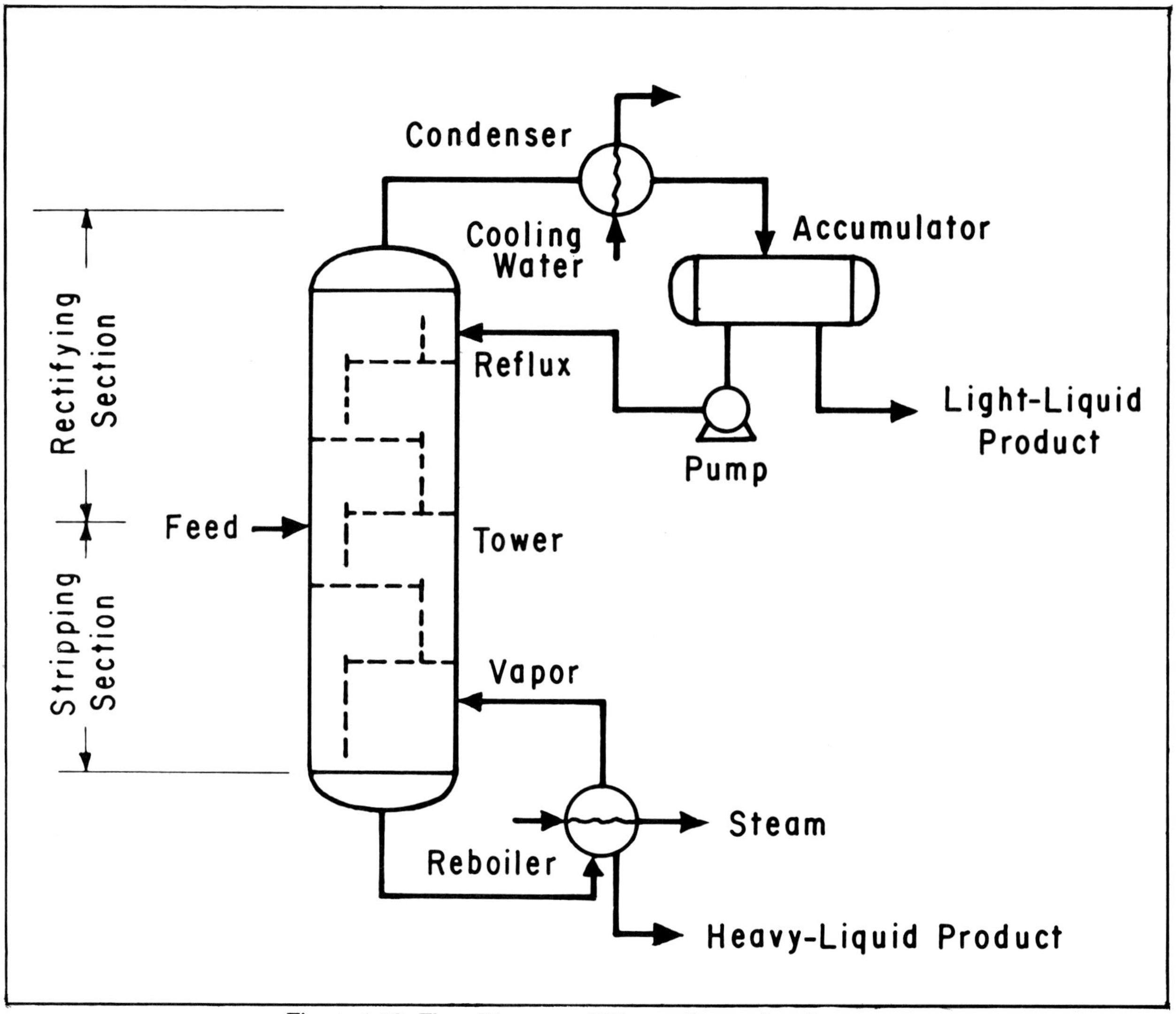

Figure 1.18. Flow Diagram of Tower Separation System

Chapter II

Oil Absorption

Oil-absorption plants are used for the removal of hydrocarbon products from natural gas. The three primary processing systems in an oil-absorption plant are recovery, rejection, and separation. These systems are illustrated in figure 2.1. Throughout this chapter, a refrigerated-absorption plant designed for ethane recovery will be used as the example plant.

Products are recovered—that is, removed—from the inlet gas in the recovery system by a combination of refrigeration and oil absorption. However, some methane is also recovered with the product. Current product specifications allow very little of the methane to remain in the plant product. The second processing system rejects the methane from the product stream. This leaves only product and absorption oil to be handled in the third processing system, separation. Each of these processing systems will be explained in greater detail in the following sections.

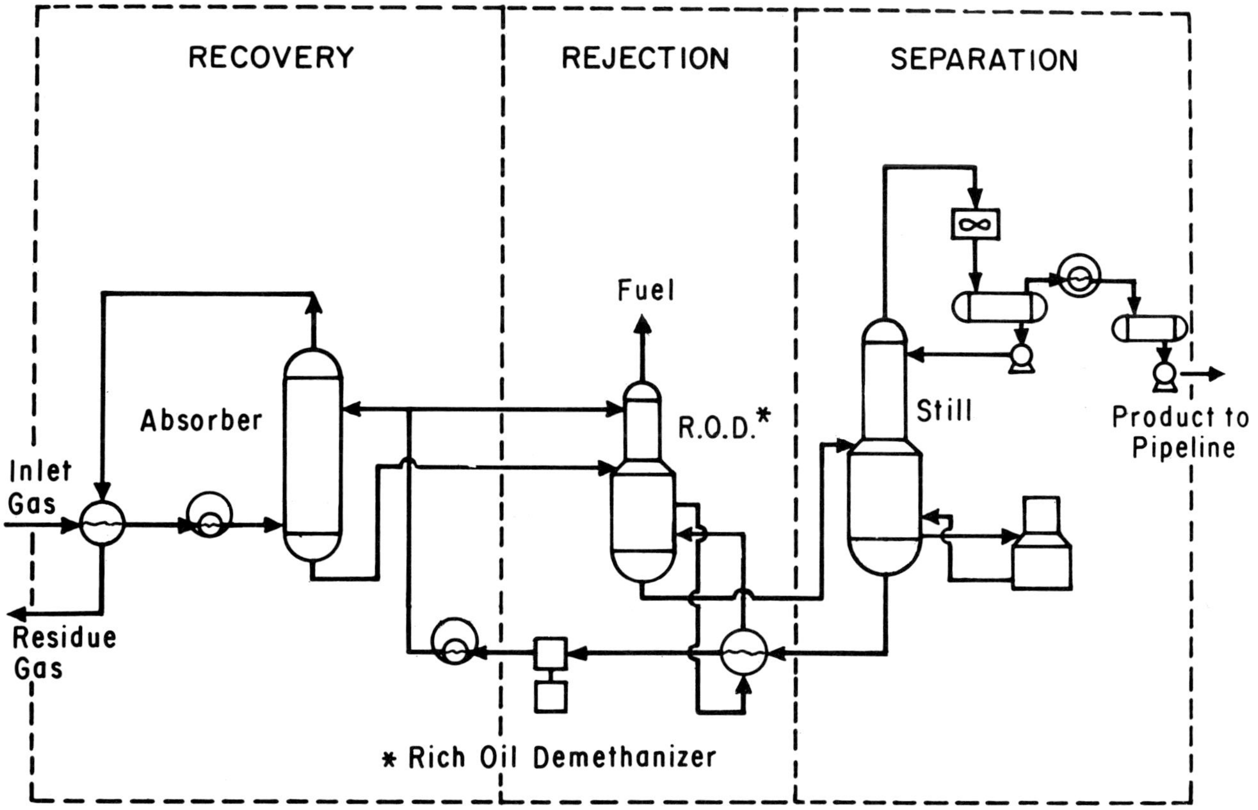

Figure 2.1. Flow Diagram of Primary Systems for Gas Processing

RECOVERY SYSTEM

All product removal from the inlet gas must be accomplished in the first processing step, which is the recovery system. The principal item of equipment is the absorber. All the other equipment in the recovery system is used to increase the operating efficiency of the absorber.

Absorption

Absorption is the process in which a liquid is used to remove some of the components from a gas. Figure 2.2 shows an absorber. Some new terms have been used in this figure. *Lean oil* is absorption oil. It is lean in the sense that it contains little or no hydrocarbon products of the type to be extracted from the gas. The physical properties of lean oil vary with each plant. However, lean oil is a liquid, the weight of which usually falls between kerosine (or a slightly heavier product) and paint thinner. In fact, most plants use a lean oil called Varsol, which is heavier than kerosine and lighter than paint thinner. The reasons for the differences in lean oil will be explained later. The liquid leaving the bottom of the absorber is called *rich oil* because it is rich in hydrocarbon products of the type to be recovered from the gas. Residue gas from an absorber is inlet gas after the product-type hydrocarbons have been removed.

In an absorber tower, the gas and oil are competing to hold the product hydrocarbons. An absorbed hydrocarbon acts like a liquid, though it was taken from the gas stream passing through the absorber unit. Hydrocarbons having lower vapor pressures (for example, gasoline and butanes) are more easily absorbed than hydrocarbons that have higher vapor pressures, such as propane, ethane, and methane. As hydrocarbon vapor pressures become higher, the hydrocarbons become more difficult to absorb because they tend to remain in the gas phase. Absorption tower efficiency is influenced by pressure, temperature, lean-oil quality, and lean-oil-to-residue-gas ratio.

Absorption is more efficient at high pressures than at low pressures. This is because the vapor pressure of high-pressure rich oil is almost the same as the vapor pressure of the hydrocarbons. This makes it easier for the hydrocarbons to be absorbed from the gas and to join the absorption oil as a liquid. Residue gas delivery pressure requirements generally set operating pressure of the absorber unit.

Absorption is more efficient at low temperatures than at high temperatures. This is because the vapor pressure of a hydrocarbon in-

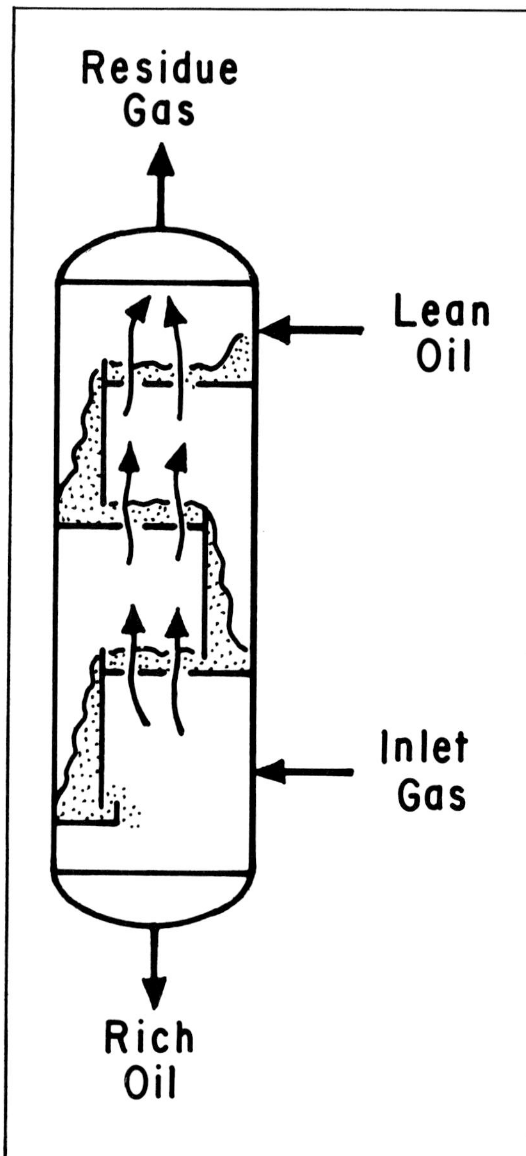

Figure 2.2. Gas and Oil Flow in an Absorber

creases with temperature. If it is desired to lower the vapor pressure of a hydrocarbon, then its temperature must be lowered. The lower the hydrocarbon vapor pressure, the easier it is for it to leave the gas and become a part of the absorption oil as a liquid.

Conventional oil-absorption plants operate at lean-oil and inlet gas temperatures that can be attained with cooling water that is near the temperature of the air around the plant. This means oil and gas temperatures are usually between 90 F and 100 F. Since about 1955, most plants that have been built use refrigeration to chill the lean oil and inlet gas to temperatures between 0 F and -40 F. Instead of cool water, a fluid such as ammonia, Freon, or propane is used for the final cooling. The majority of these plants use propane as the refrigerant.

Figure 2.3 is a schematic flow diagram of a refrigerated-absorber system. Inlet gas is first cooled by the absorber residue gas in the gas/gas (gas-to-gas) exchanger. Final cooling of the inlet gas occurs in the gas chiller. Figure 2.4 shows a gas chiller. The warm gas flows through the tubes of the chiller while the refrigerant is in the chiller shell surrounding the tubes. Chilled lean oil is fed to the top of the absorber.

Absorption is actually the transfer of molecules from the gas phase of the inlet gas to the liquid phase, which is the rich oil. The capacity of the rich oil at a certain pressure and temperature to absorb gas molecules depends on the number of lean-oil molecules. This being the case, as many molecules of lean oil as possible must be used, and this is sometimes easier said than done.

Engineering design must consider that all of the circulated lean oil must be heated to the highest temperature at the still bottoms and

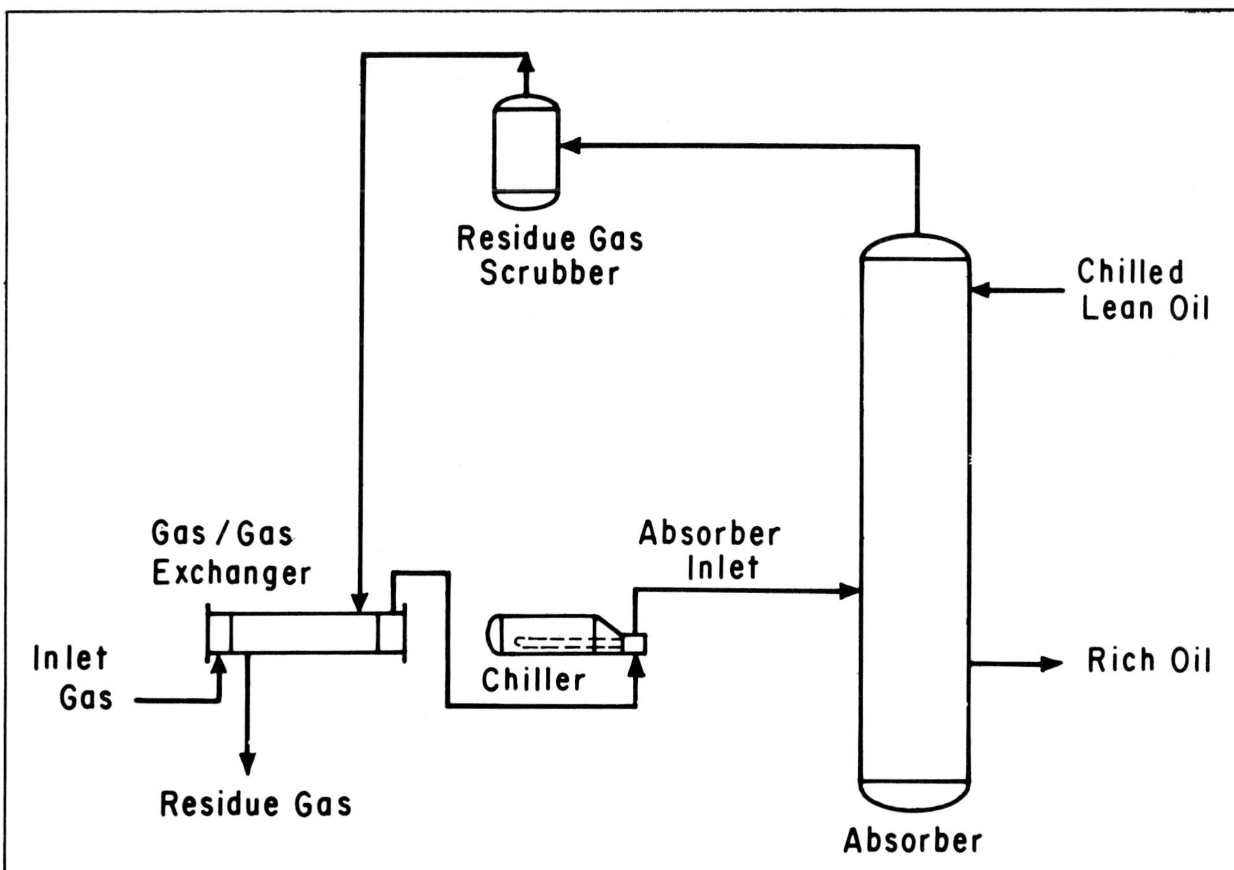

Figure 2.3. Flow Diagram of Refrigerated Absorber

cooled to the lowest temperature at the absorber inlet. The amount of heating and cooling depends on the quantity, or pounds, of lean oil. This means that the lean oil should have a small number of pounds for each mole. The number of pounds per mole is called the molecular weight.

Figure 2.4. Gas Chiller

Molecular weight is not an absolute number; rather, it is a relative measure of the weight of one substance as compared to the weight of a standard substance. Molecular weight is a measure similar to the measure that is often used to express the very high speeds that some jet airplanes are capable of attaining. This measure of speed is called a Mach number. A Mach number is the speed of the aircraft relative to the speed of sound at standard conditions of air temperature and pressure. Similarly, molecular weight is also a relative number.

As stated before, a lean oil with a low molecular weight should be employed. The lower the molecular weight, the lighter the lean oil. However, one complication with lighter lean oil is that more of it is lost by vaporization into the residue gas than that which is lost by a heavier lean oil. This being the case, each plant has a lean oil best suited for its operating conditions. The lean oil selected is the one that has the most molecules available for absorption while at the same time has characteristics that keep vaporization losses at an economical level.

Once the lean-oil molecular weight for a specific plant has been established, the oil-to-gas ratio can be used to express how hard the absorber is being worked. The oil-to-gas ratio is usually expressed as the gallons per minute (GPM) of lean oil circulated per million cubic feet of gas passed through the absorber. One million cubic feet per day is often abbreviated to MMSCFD, which means million standard cubic feet per day; i.e., gas volume measured at 60F and 14.7 psia. This expression is practical because the gallons of lean-oil flow can be measured and the gas flow in cubic feet through the plant is known. It can be seen from figure 2.5 that the percentage recovered in the absorber of both ethane and propane increases as the lean-oil-to-inlet-gas ratio is increased. The reason for this is that more lean-oil molecules are available for absorption of the gas molecules at the higher lean-oil-to-inlet-gas ratios.

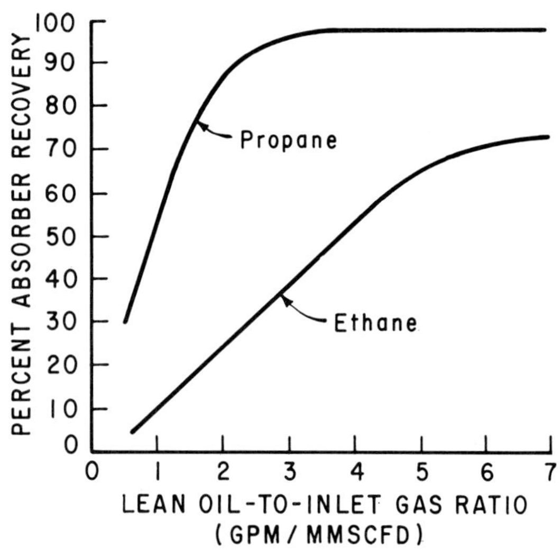

Figure 2.5. Effect of Lean Oil-to-Inlet Gas Ratio on Absorber Recovery

Presaturation

As a product is absorbed from the gas and becomes a liquid with the rich oil, the latent heat is given up to the oil and gas. This increases the temperature of both streams within the absorber and lowers the absorber efficiency because of the higher temperatures. Various methods are used to remove some of this heat of absorption outside of the absorber, thereby minimizing the temperature increase inside the absorber. One method frequently used is *presaturation.* Presaturation is not essential to the process, but, without presaturation, more lean oil has to be circulated to achieve the same product recovery.

Figure 2.6 shows a low-pressure presaturation system using the low-pressure vapors from a rich-oil demethanizer, or ROD. (RODs will be discussed later.) These low-pressure vapors are mixed with the lean oil and then chilled by the plant refrigerant. The vapors not absorbed by the lean oil are separated from the presaturated lean oil in the accumulator (fig. 2.7). These low-pressure vapors are essentially methane. However, presaturation is effective because a large part of the hydrocarbons removed from the gas by the lean oil in the absorber is methane.

Figure 2.7. Presaturator-Accumulator

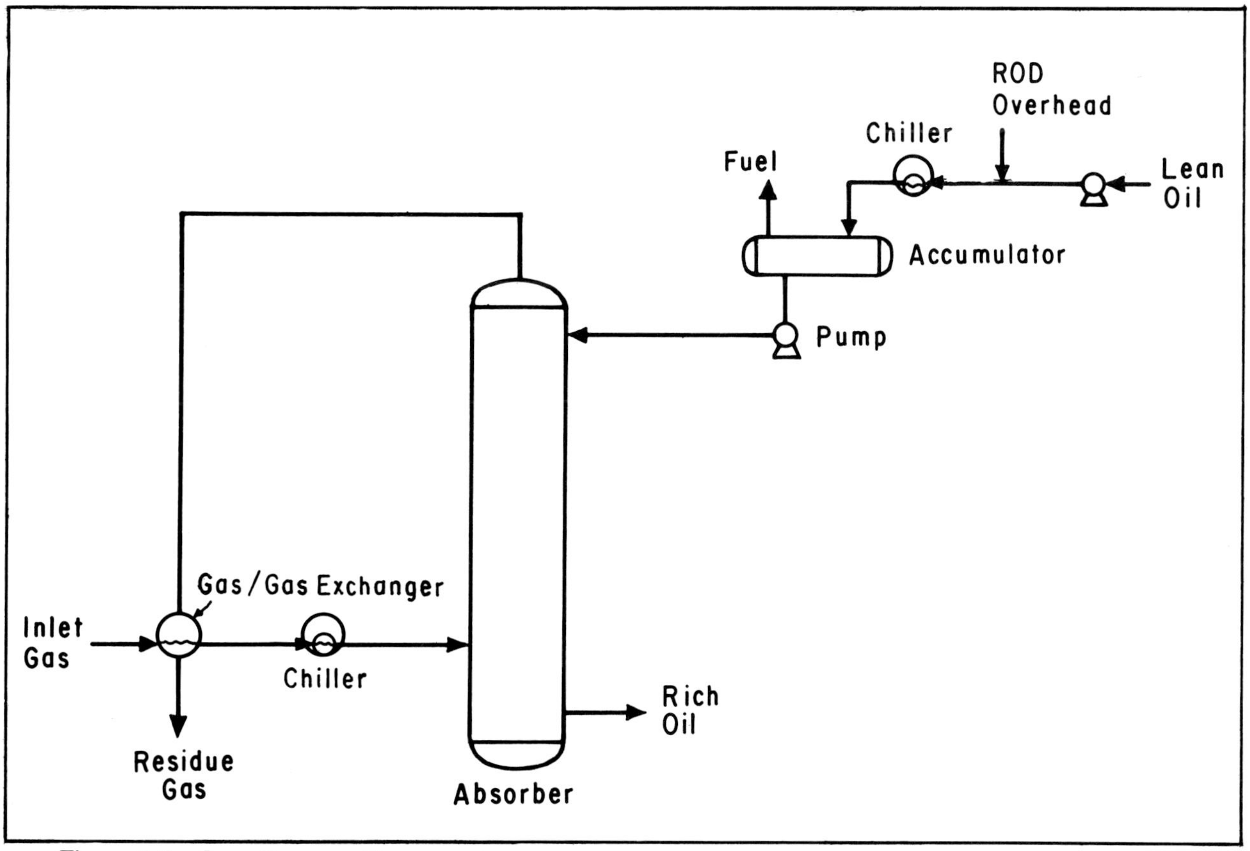

Figure 2.6. Low-Pressure Presaturation System Using Vapors from Rich-Oil Demethanizer (ROD)

Possible Problems

Usually problems with the recovery system are due either to absorber flooding or freezing. Figure 2.8 shows a refrigerated absorber with the desired instruments and controls for quick detection of upsets.

When an absorber floods, some (if not all) of the lean oil fed to the absorber leaves the top of the absorber with the residue gas. As can be seen from the flow diagram in figure 2.8, the lean-oil carry-over from the absorber will be caught by the absorber residue scrubber. The high-liquid-level alarm (HLA) installed on the residue gas scrubber will alert the operator when lean oil is being carried over.

Many plants have a differential pressure indicator (DPI) to measure the pressure drop across the absorber. When the absorber floods, there is a large increase in the differential pressure. The DPI detects absorber flooding before the high-level alarm on the residue gas scrubber.

Three things can cause an absorber to flood: (1) a sudden, large increase in either gas or lean-oil flow; (2) too much lean-oil flow, even if it is gradually increased, for the gas volume being handled; and (3) plugging of gas or lean-oil flow paths due to dirt or freezing.

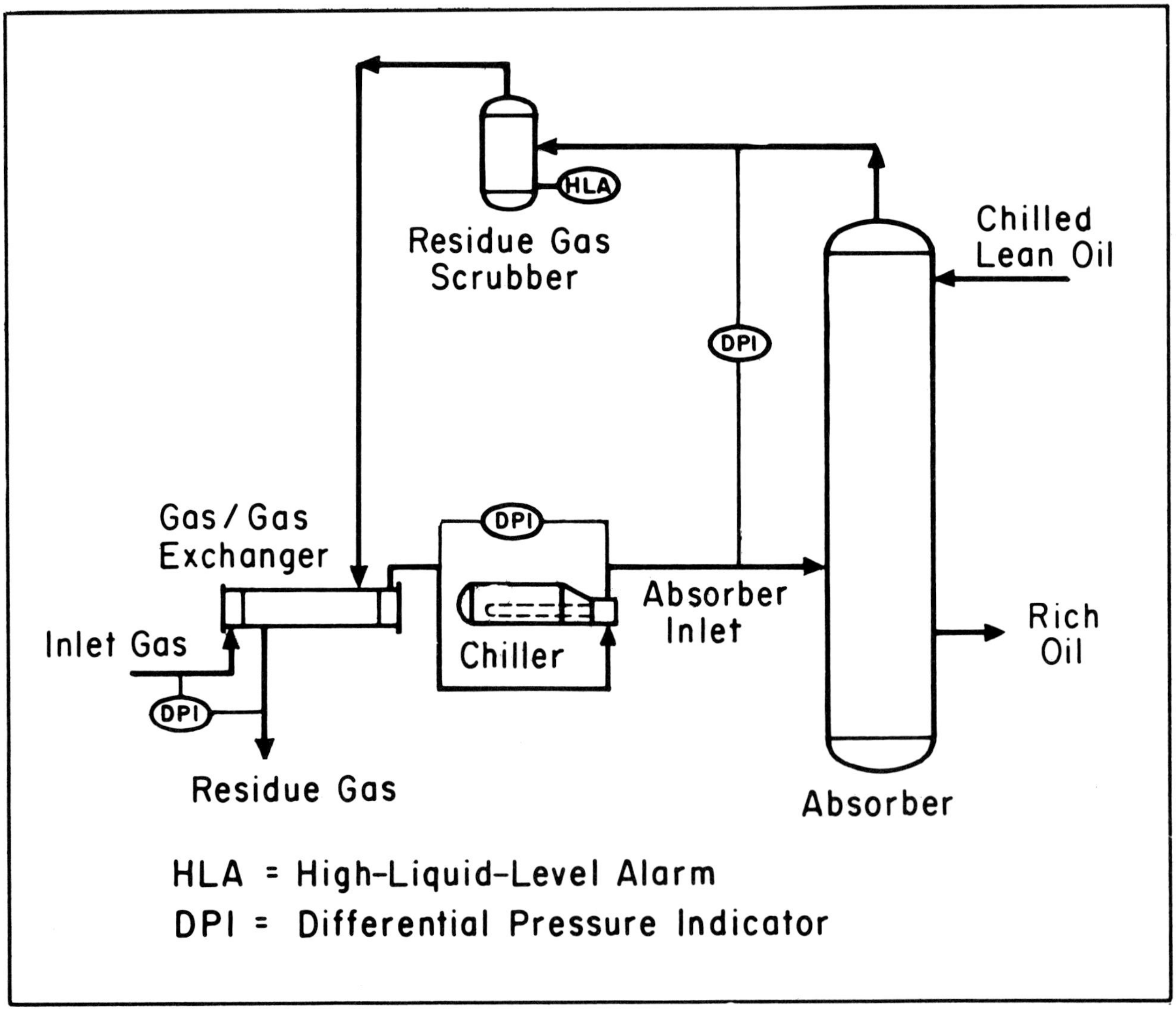

Figure 2.8. Refrigerated Absorber Showing Instruments

When a sudden, large increase in gas volume must be handled by an absorber that is near its capacity, such as that caused by putting a large inlet gas compressor on the line, the lean-oil flow to the absorber should be reduced. After the absorber has had time to "settle" from the increase in gas volume, the lean-oil flow may be gradually increased.

The inlet gas and lean-oil capacities of an absorber are interdependent. Figure 2.9 shows that the absorber's capacity for lean-oil flow decreases as the inlet gas volume increases. This absorber has a lean-oil flow capacity of 1,000 gal/min when the inlet gas volume is 200 million cubic feet per day (MMSCFD). At 250 MMSCFD, the lean-oil flow cannot be greater than 800 gal/min or flooding will occur. When an absorber floods, a large reduction must be made in the lean-oil flow to the absorber. This will quickly stop the flooding. Then the cause of the flooding should be determined.

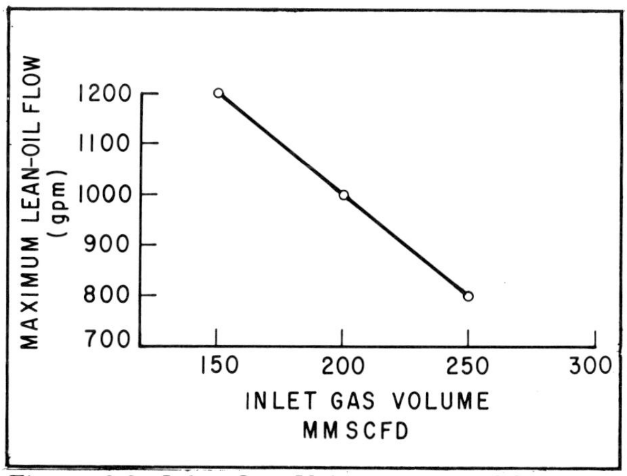

Figure 2.9. Inlet Gas Volume versus Maximum Lean-Oil Flow

When freezing occurs, part of the normally available flow space is occupied by hydrates. This causes an increase in the pressure drop across the equipment. Figure 2.8 showed several differential pressure indicators (DPIs) located so that an operator may quickly detect freezing and its location. Freezing in the inlet gas train will also increase the gas temperature to the absorber. When freezing occurs, injection of methanol on a temporary basis may be effective in preventing additional freezing and will break up the hydrates already formed. The only solution to a bad freeze-up is to allow the frozen area to warm up.

REJECTION SYSTEM

A rejection system is necessary to get rid of the methane that has been retained by the recovery system and that cannot be sold as a liquid product. In other words, the product must be demethanized. Usually, more than 50 percent of the hydrocarbons removed from the inlet gas is methane. It is easy to figure out ways to reject the methane from the rich oil—just do the opposite of those things that help absorption; that is, reduce pressure and increase temperature. While methane is being rejected, however, it is important to keep the good product in the rich oil. This means that demethanization is a combination of rejection and reabsorbing.

Most plants accomplish methane rejection in two steps. The primary methane rejection equipment, such as a rich-oil flash tank, partial demethanizer, reboiled absorber, and so forth, usually eliminates at least half the methane from the rich oil. This methane is rejected by reducing pressure, increasing temperature, or by a combination of pressure reduction and temperature increase. The rejected methane stream usually has sufficient ethane and propane to make reabsorption desirable.

Hot Rich-Oil Flash Tank

There are many combinations of rejection and reabsorption. Figure 2.10 shows a hot rich-oil flash tank being used for methane rejection. Methane is rejected by adding heat to the absorber rich oil upstream of the rich-oil flash tank. The absorber rich oil is pumped to the higher pressure of the rich-oil flash tank. The higher pressure allows the flashed vapor to be recycled to the inlet gas absorber without being recompressed. The inlet gas absorber is also the reabsorber for the rich-oil flash tank vapor.

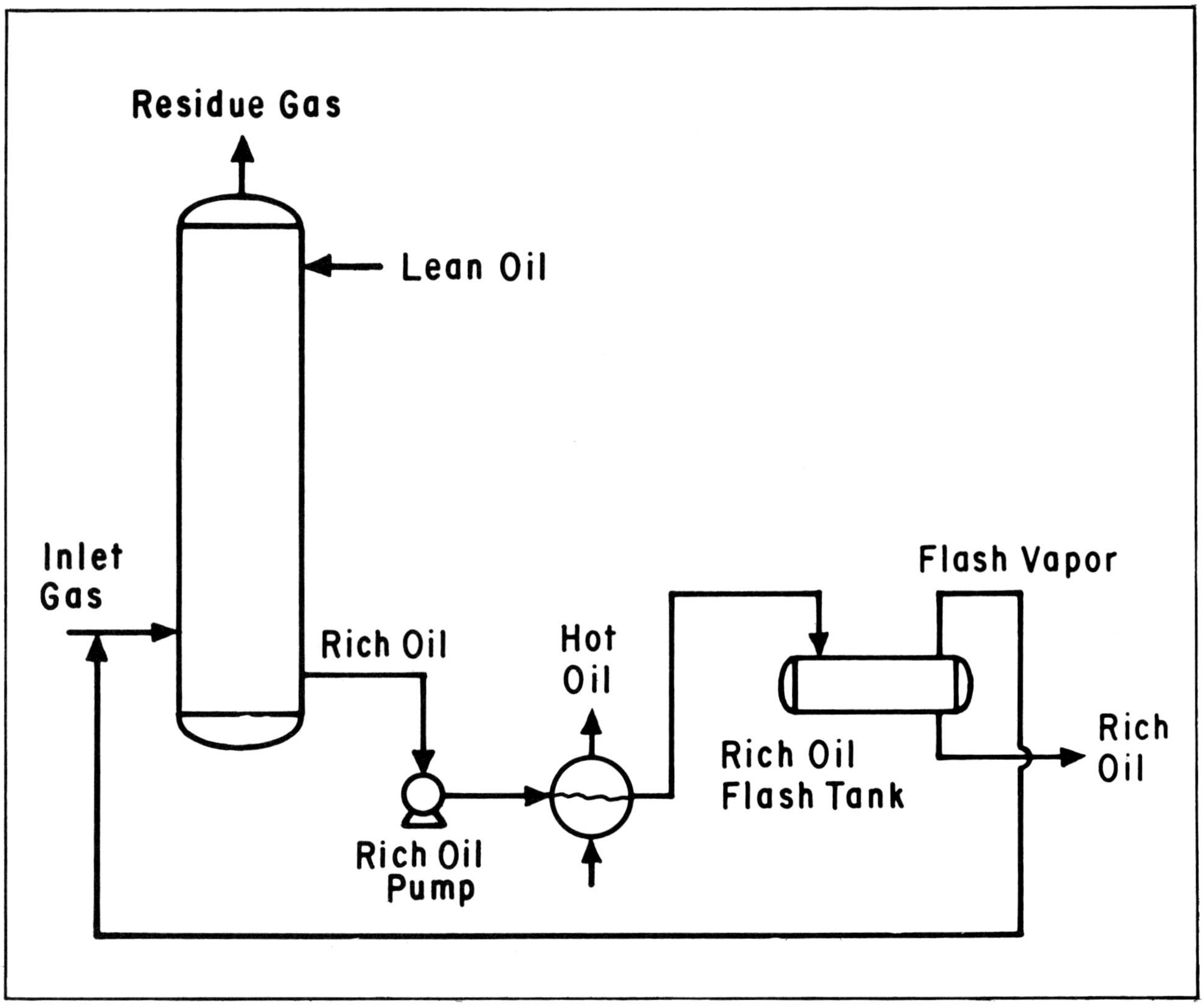

Figure 2.10. Hot Rich-Oil Flash Tank Arranged for Methane Rejection

Rich-Oil Demethanizer

The heart of the rejection system is the rich-oil demethanizer (ROD), as shown in figure 2.11. This is the second and final step for methane rejection. The rejected methane leaves the top of the rich-oil demethanizer, while the product and lean oil leave the bottom. A rich-oil demethanizer allows rejection of the rich-oil methane, while retaining most of the ethane. Figure 2.12 is a flow diagram for a rich-oil demethanizer. The feed may come from a rich-oil flash tank, a partial demethanizer, a reboiled absorber, or straight from the main absorber. Heat is added either to the total feed or to most of the feed, while the remainder of the feed (which is cold rich oil) is used similarly to lean oil. The tower performs two functions. The bottom section *strips*—that is, it eliminates by heat and vapor flow—the methane from the rich oil. The top section reabsorbs the ethane and other products from the vapor leaving the bottom section.

Lean oil is used in the top section just as lean oil is used in an absorber. A reboiler is used in the bottom section to supply the heat and vapor necessary for good stripping. The top-section lean-oil feed controls the amount of ethane lost in the overhead vapor. The temperature of the rich oil leaving the stripping section, called the *rich-oil demethanizer bottoms,* controls the methane content of the rich oil.

Changes in plant operating conditions may require changes in the rich-oil demethanizer's bottom temperature. The pressure of the rich-oil demethanizer is controlled and should be held constant. The vapor pressure of the rich-oil demethanizer bottoms is equal to the tower

Figure 2.11. Rich-Oil Demethanizer (ROD)

operating pressure. Therefore, the temperature of the bottoms controls its composition. The proper bottoms composition is usually expressed as the ratio of mole percent methane to mole percent ethane.

An *increase* in the bottom temperature *decreases* the methane-to-ethane ratio. A *decrease* in the bottom temperature *increases* the methane-to-ethane ratio. Think of the bottoms as being two components: product and lean oil. Product is lighter, thus it has a higher vapor pressure than lean oil. This means that it takes a higher temperature to raise the lean-oil vapor pressure to the tower operating pressure than it takes to raise the product vapor pressure to the tower operating pressure. If the bottom temperature is too high, too much product is driven overhead. If the bottom temperature is too low, too much methane is left in the bottoms.

The proper way to think of bottom-temperature adjustments is to think of the lean-oil-to-product ratio in the bottoms. A high ratio means a higher bottom temperature because it requires a higher temperature to raise the lean-oil vapor pressure. A low ratio means a lower bottom temperature. Figure 2.13 is a graph of demethanizer bottom temperature versus lean-oil-to-product ratio. This graph was drawn from actual plant performance tests. While it is not sufficiently accurate to be used for general plant operations, it does show the relationship of bottom temperature and the lean-oil-to-product ratio.

Plant operating changes that change the lean-oil-to-product ratio require changes in the rich-oil demethanizer bottom temperature. Figure 2.14 shows the bottom temperature adjustments required for plant operating changes. In reviewing this, think of what is happening to the lean-oil-to-product ratio. For example, if there is a *decrease* in inlet gas volume, the plant will recover less product. Therefore, there will be an *increase* in the ratio of lean oil to product in the rich-oil demethanizer bottoms. This requires an *increase* in the rich-oil demethanizer bottom temperature to reject the proper amount of methane from the rich oil.

Possible Problems

The rejected methane from the ROD can be used to presaturate the lean oil, can be consumed as plant fuel, or can be recompressed to join the inlet gas or absorber residue gas. Most plants use a combination of these methods. However, any system has an ultimate capacity for the volume of rejected methane that it can handle. When this capacity is reached, then the quantity of methane fed to the ROD must be reduced. This can be accomplished by increasing the rejection of methane upstream of the ROD or by reducing the lean-oil flow to the absorber, which will decrease the methane recovery.

Improper demethanization of the product will not only make a product of improper specification, but often it will cause excess pressures in the still and product-condensing facilities. Proper demethanization of the product is primarily dependent on the bottom temperature of the ROD. In many plants, the heat supply for the ROD reboiler is the hot lean oil from the bottom of the still. This is illustrated in figure 2.15. When the available heat input from

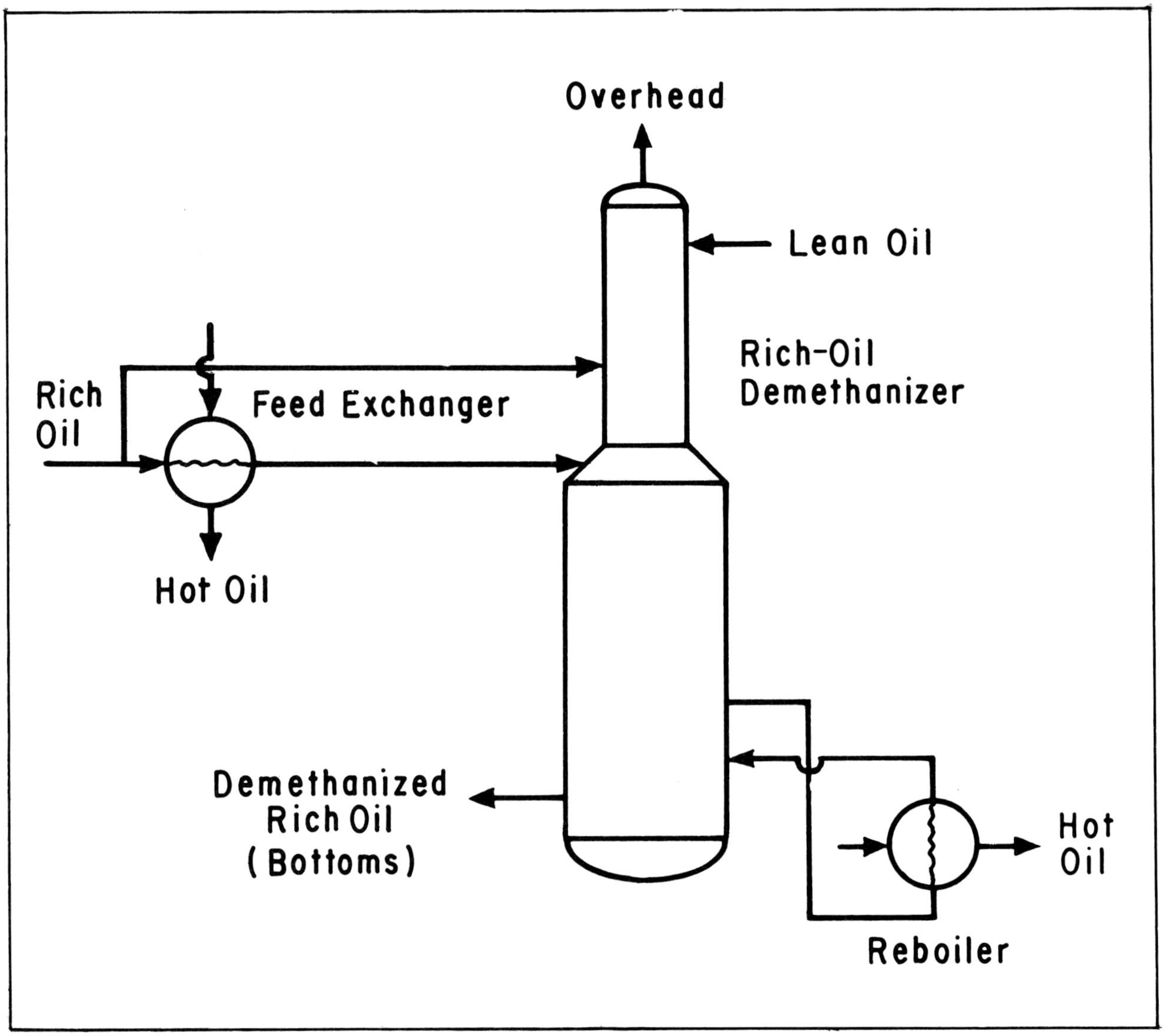

Figure 2.12. Flow Diagram for ROD

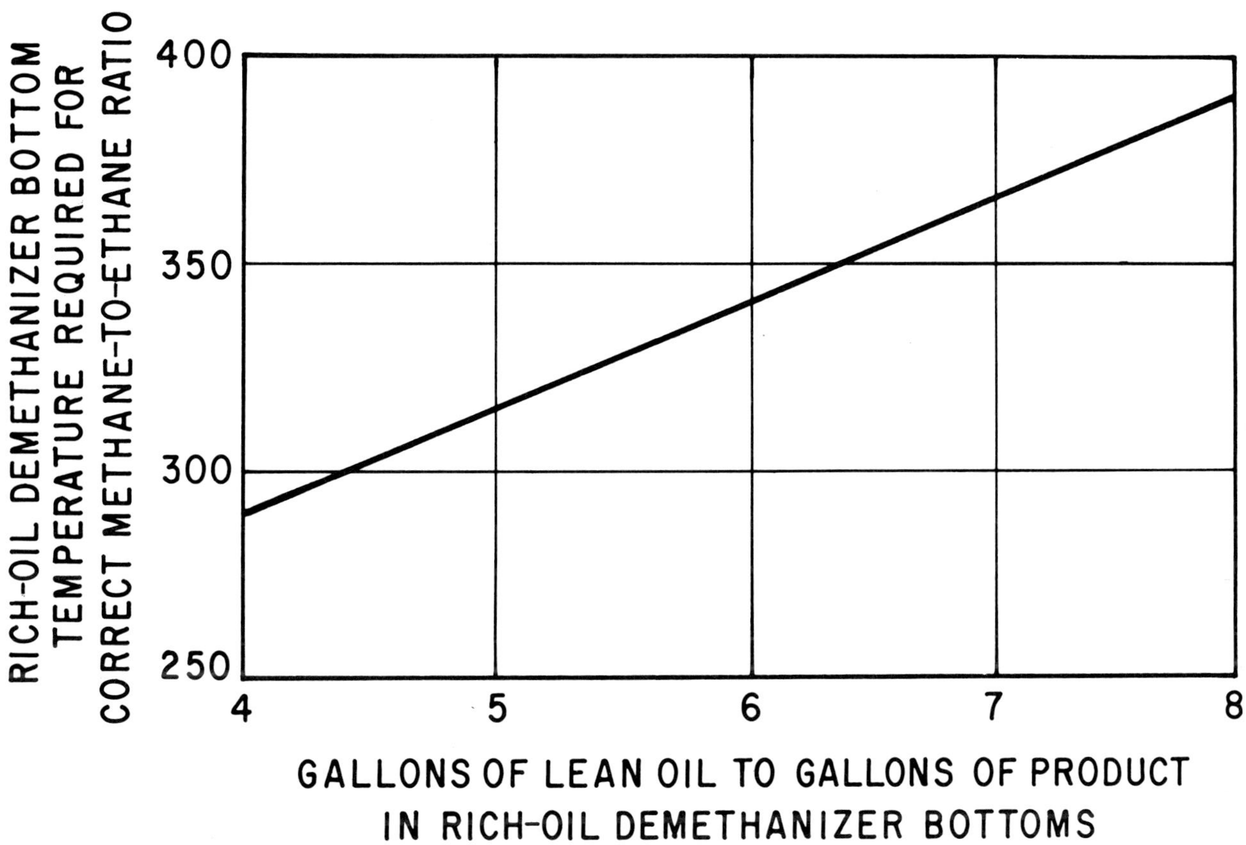

Figure 2.13. Required ROD Bottom Temperature

Reasons for Making Adjustment	
1. Maintain methane content specification in product 2. Recover maximum amount of ethane	
Operating Condition Change	**Adjustment**
Decrease in inlet gas rate	Increase
Increase in inlet gas rate	Decrease
Decrease in lean-oil rate	Decrease
Increase in lean-oil rate	Increase
Decrease in processing temperature	Decrease
Increase in processing temperature	Increase
Refrigeration compressor down	Increase

Figure 2.14. ROD bottom-temperature adjustments must relate to changes in plant operation.

the hot lean oil is inadequate, then the quantity of methane fed to the ROD must be reduced.

SEPARATION SYSTEM

The ROD bottoms consist of product and lean oil. This stream is fed to the third processing system of the example plant, the separation system. Here, the product is separated from the lean oil. The principal piece of equipment in the separation system is the still, sometimes called a rich-oil fractionator. The product goes out the top of the still, and the lean oil leaves the bottom of the still. This method of separation is frequently called *distillation.*

Distillation is essentially the reverse of absorption. This means that distillation works best at high temperatures and low pressure. For efficient distillation, the operating conditions must make it easy for the product to leave the liquid rich oil and become a vapor. Removal of the product from the rich oil requires a stripping

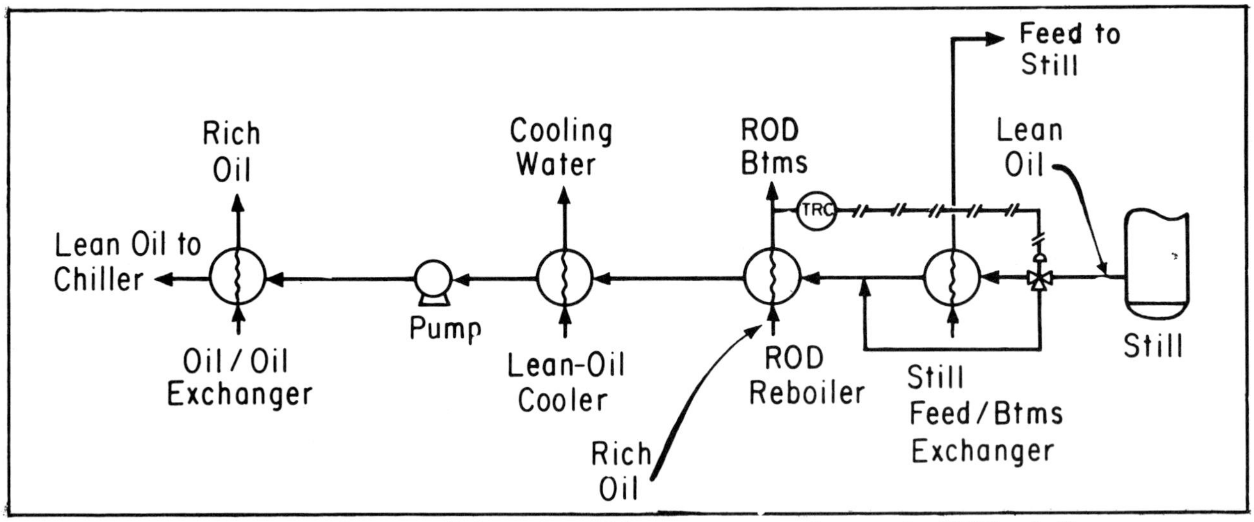

Figure 2.15. Hot lean oil from the still bottom supplies heat to the ROD reboiler.

medium in the vapor form. Steam is frequently used, and stills using steam are sometimes called *wet stills.* Hydrocarbon vapors can also be used as the stripping medium, and these stills are called *dry stills.* Dry distillation is used to keep the lean oil free of water. Dry lean oil is necessary in refrigerated plants to avoid freezing or hydrate problems when the oil is chilled.

Still

A flow diagram of a dry still is shown in figure 2.16. Hydrocarbon vapors are made by partial vaporization of the hot oil pumped through the still reboiler. These hydrocarbon vapors are used as the stripping medium to separate the product from the lean oil. Reflux is used to keep lean oil from going out the top of the still with the product. In the flow diagram of a dry still, the still reboiler is shown as a fired heater. Plants with light lean oils, which have low molecular weight, often use either high-pressure steam or another heating medium in a shell-and-tube exchanger for the still reboiler. For this type of system, the hot steam to the reboiler is controlled to maintain the oil temperature.

Operation of the still is generally controlled by using a constant reflux quantity and supplying sufficient heat to the reboiler to maintain a fixed lean-oil outlet temperature. A still controls the 50 percent boiling point of the lean oil and the final boiling point of the product. Therefore, still performance can be and should be checked by running American Society for Testing Materials (ASTM) distillation tests on

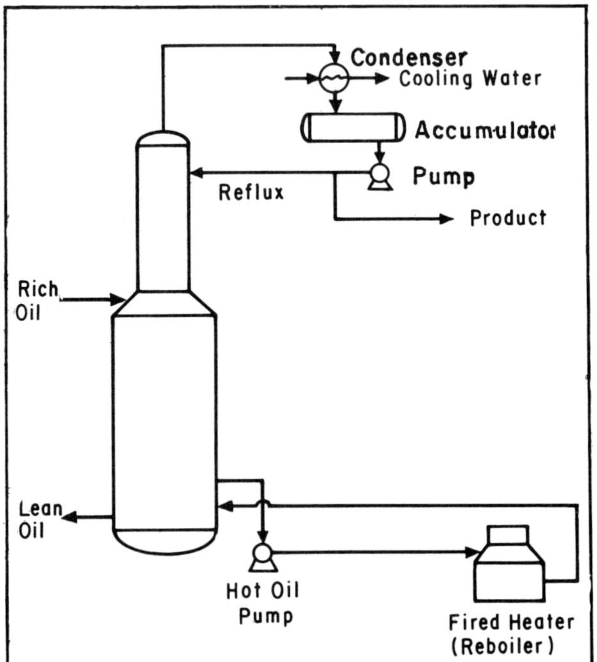

Figure 2.16. Dry-Still Processing System

the lean oil and the product. The product sample should be allowed to "weather" before running the ASTM distillation test. The primary interest to the operator is the final boiling point temperature of the product sample. This is not the "dry-point" temperature that is observed at the end of an ASTM distillation test. The final boiling point is the observed temperature during an ASTM test when the temperature levels off as the last of the test liquid boils off. The dry point is the observed temperature as the final residue of the test dries out. A complete ASTM distillation test should be run on the lean oil to assess its quality.

Good lean-oil quality is essential to efficient plant operation. The principal standards of lean-oil quality are the 50 percent boiling point temperature and the distillation range of temperatures. When these two standards are correct, then the molecular weight of the lean oil is fixed. At most plants, maintaining good lean-oil quality requires efficient operation of both the still and the oil purifier.

Oil Purification

Probably the most neglected system in a gas processing plant is the oil purification, or oil reclaimer, system. Oil reclaiming is necessary to control the final boiling point temperature of the lean oil. This is accomplished by removing heavy ends from the plant lean oil. The presence of heavy components in lean oil will reduce product recovery, increase operating difficulties, and increase lean-oil losses.

There is almost nothing standard in an oil reclaiming system. Some are batch operated, and some have continuous feed with automatic operation. Figure 2.17 shows one type of oil reclaiming system. Even though the operation varies with almost each installation, the primary purpose of all oil reclaimers is the same: to control the final boiling point temperature of the lean oil. This can only be accomplished by not letting heavy ends go overhead from the oil reclaimer still.

The curves in the graph (fig. 2.18) illustrate why it is necessary to control the final boiling point temperature of the lean oil. On the right side of the graph, it can be seen that the bad lean oil has a high final boiling point temperature (FBP). As heavy ends build up and form a "tail," the oil tries to balance itself by holding the light ends as a "head." This can be seen on the left side of the graph by the low initial boiling point temperature (IBP) of the bad lean oil. Because of the wide range of distillation temperatures, operations at the plant from which this example was made were very difficult until the heavy ends were removed from the lean oil by reducing the overhead temperature of the oil reclaimer. The result is shown by the curve illustrating the good lean oil that is within the desirable distillation range of 100 F between initial and final boiling points.

Possible Problems

Leaving product hydrocarbons in the bottom of the still with the lean oil must be avoided because the product will be vaporized and leave the absorber with the residue gas when the lean oil is pumped to the absorber. The initial boiling point of the lean oil is a good measure of product separation efficiency. Figure 2.19 shows the product losses experienced by one plant, compared to the proper intitial boiling point temperature of the lean oil. At a lean-oil initial boiling point temperature of 75 F less than design, the loss of products amounted to 4.4 gallons per million standard cubic feet (MMSCF) of absorber residue gas.

The final boiling point temperature of the product is a measure of how much lean oil is leaving the top of the still with the product. If lean-oil "make-up" is being purchased, then the lean oil in the product can be minimized by pumping sufficient reflux to the still. One indication of sufficient reflux is that the final boiling point temperature of the product should not be higher than the initial boiling point temperature of the lean oil.

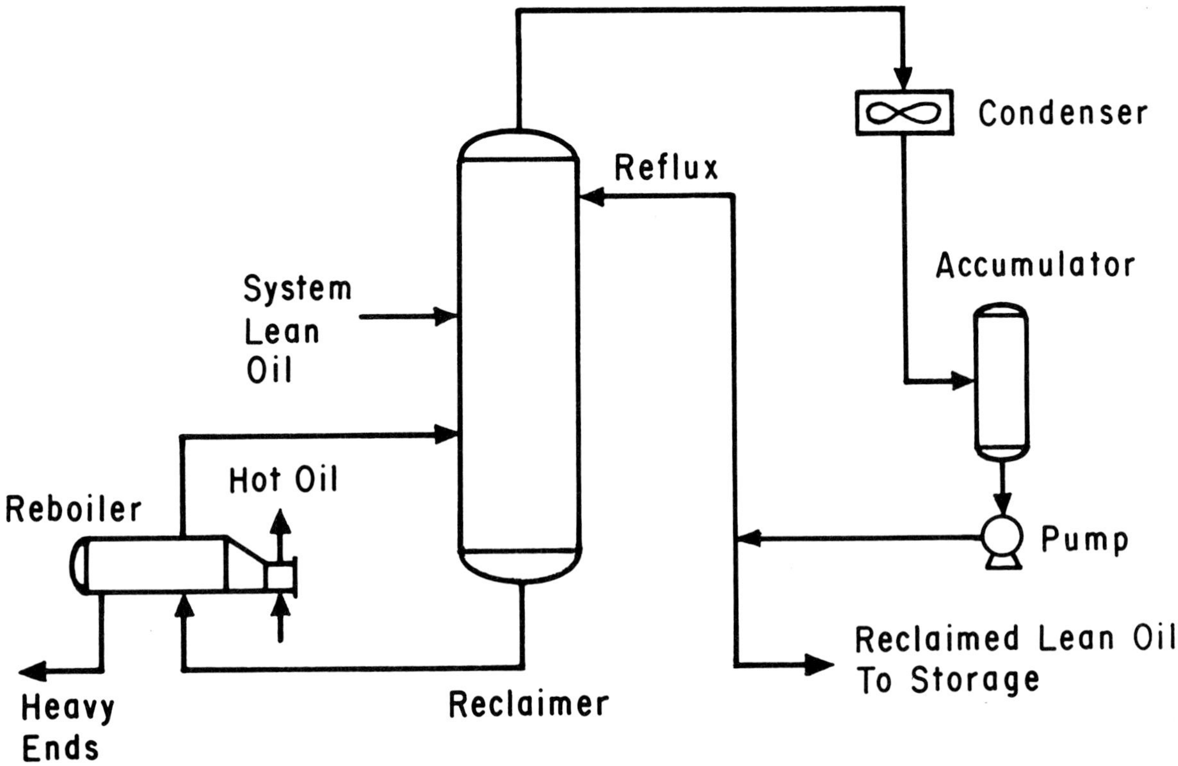

Figure 2.17. Oil Reclaimer

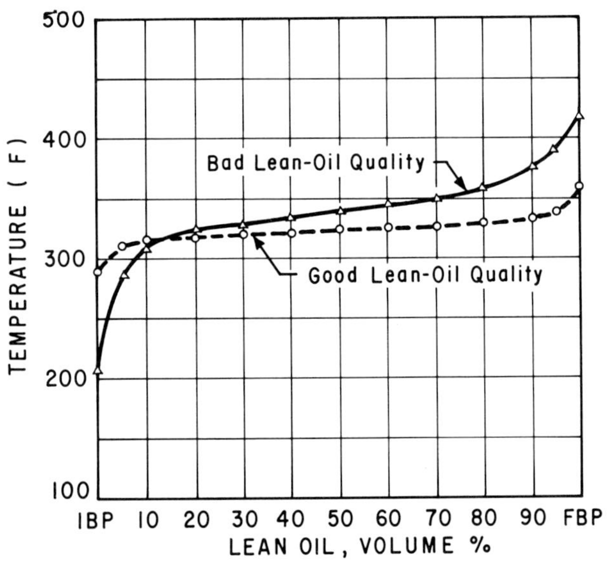

Figure 2.18. Distillation tests show lean-oil quality.

If a multipass fired heater is used for the still reboiler, it is essential that the oil flow and outlet temperatures for each pass be balanced. If either the flow or outlet temperature becomes very much out of balance, there is the danger that some of the heater tubes will become *dry*—that is, liquid will not be flowing through the tubes. Dry tubes quickly "coke" and form hot spots. These hot spots soon develop into tube failures. For this reason, it is very important to maintain a high circulation rate to a fired heater. Since the amount of oil vaporized is relatively constant, then a high circulation rate will insure that enough oil will remain a liquid at the heater outlet to help avoid hot spots.

REFRIGERATION SYSTEM

An auxiliary system essential to a refrigerated-absorption plant is the refrigeration system. To achieve cold processing temperatures, most plants use propane as a refrigerant. A typical propane refrigeration system is shown in the flow diagram in figure 2.20. A plant refrigeration system is very similar in operation to a

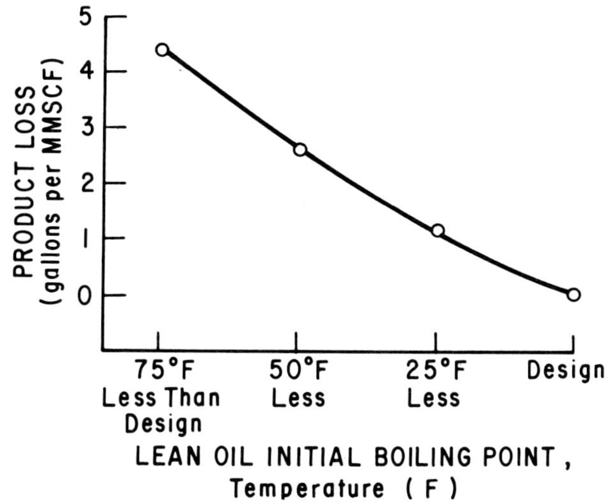

Figure 2.19. Product Losses Related to Lean-Oil Initial Boiling Point

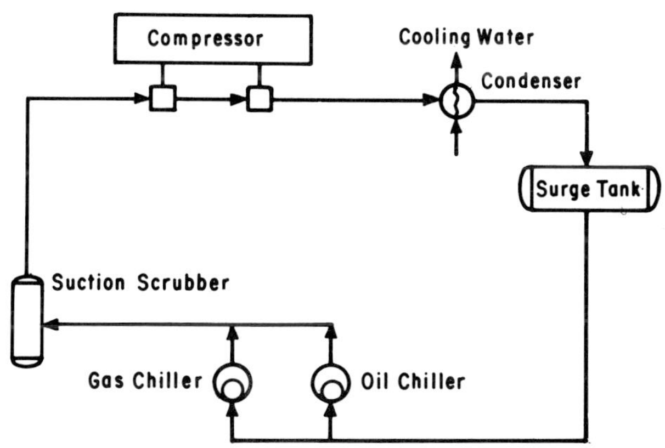

Figure 2.20. Typical Propane Refrigeration System

house air conditioner, except that an air conditioner usually uses Freon as a refrigerant. In a plant refrigeration system, propane flows as a liquid from the surge tank to the oil chiller and the gas chiller. In the house air conditioner, the chiller is called an evaporator. As the hot air is blown across the evaporator, the refrigerant is vaporized. In a plant, the oil and gas are hotter than the propane refrigerant. As the oil and gas flow through the tubes of the chillers, the propane that is in the chiller shell and covers the tubes is vaporized.

The propane vapor from the chillers flows through the suction scrubber and to the compressor. The compressor is needed in order to raise the pressure of the propane vapor from the chillers so that the propane may be condensed when it is cooled, just as Freon is compressed and condensed in a house air-conditioning system. When propane vapor is compressed, it can be condensed with cooling water or air coolers and returned to the surge tank as a liquid.

Economizers

A propane refrigeration system depends on the recovery of latent heat (which is the heat necessary to change propane from a liquid to a vapor) in the chillers. This means that propane must go into the chillers as a liquid. The propane in the surge tank may be thought of as the liquids from a high-pressure oil well. When the oil is produced from the well and flows into a stock tank, some of the vapor from the oil will leave the top of the tank, and the liquid left is the produced oil. One way to increase oil recovery in the stock tank is to produce the well through several separators in a series—that is, use stage separation. Simple oil-well producing arrangements are shown in figure 2.21. An arrangement similar to stage separation can usefully be employed in a propane refrigeration system of a gas processing plant.

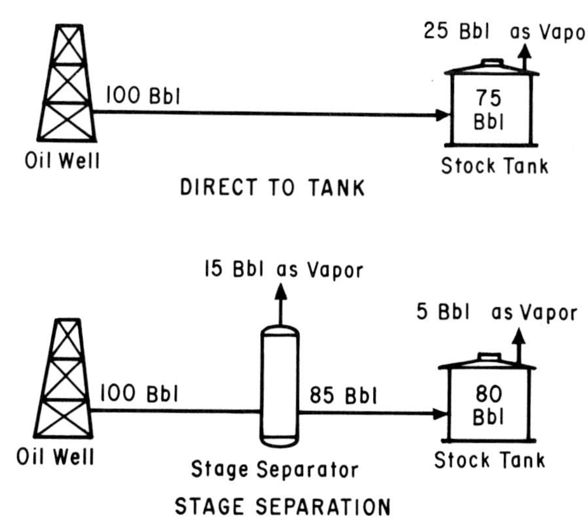

Figure 2.21. Oil Flow from Well to Stock Tank

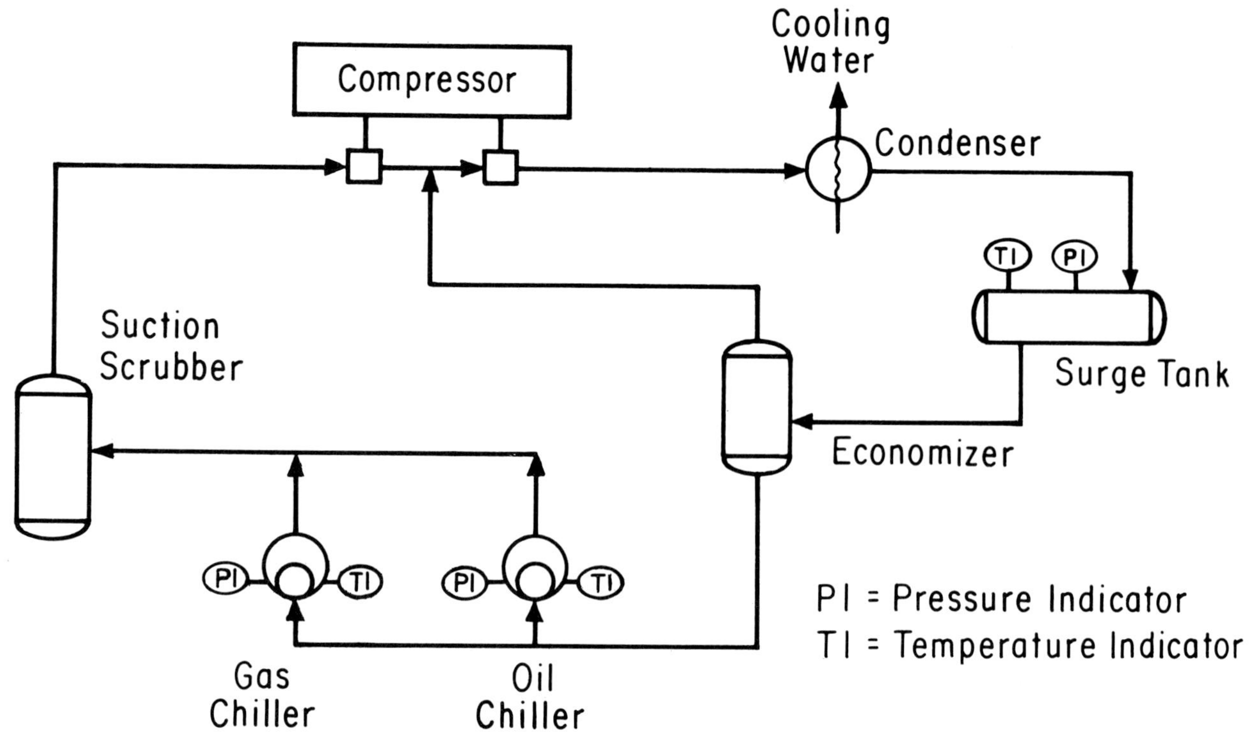

Figure 2.22. Economizer System

Figure 2.23. Two Economizers

Figure 2.22 illustrates a refrigeration system utilizing stage separation. The only difference is that gas processors call the stage separator an *economizer* because it saves liquid when it is used in a refrigeration system. An economizer not only enables more of the surge tank propane to reach the chillers as a liquid, but it also saves compressor horsepower by letting some of the propane vapor go interstage to the compressor. Some refrigeration systems have several economizers, each at a different pressure. These different pressures correspond to compressor interstage pressures. The two economizers of one refrigeration system are shown in figure 2.23.

Chillers

The gas or lean oil flows through all of the chiller tubes. To get maximum refrigeration, all of the tubes must be covered with propane. Space must be left at the top of the chiller shell to allow the vapor to separate from the liquid and flow to the compressor without carrying liquid with it. A chiller is shown in figure 2.24. The level controller is set to maintain the liquid level just above the top of the tubes. As the li-

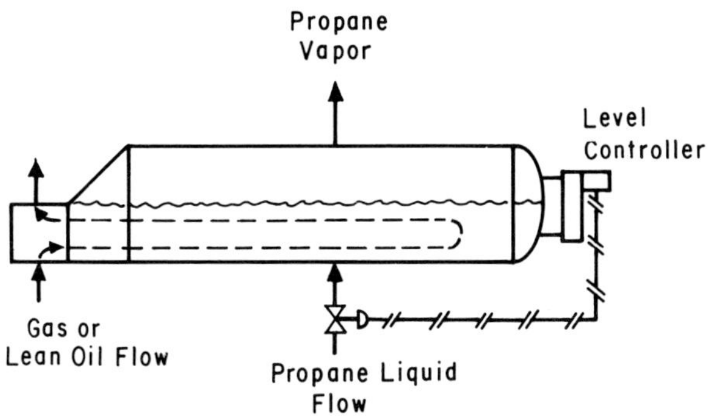

Figure 2.24. A level controller maintains the liquid level of the propane above the tubes in a chiller.

quid is vaporized, the level controller opens the control valve and lets more propane liquid flow into the chiller. With steady flows of oil and gas, the level controller will have the control valve open just enough to keep a steady flow of propane liquid coming in to hold the proper liquid levels in the oil chiller and gas chiller.

Possible Problems

The propane used in plant refrigeration systems is not pure propane. It contains some

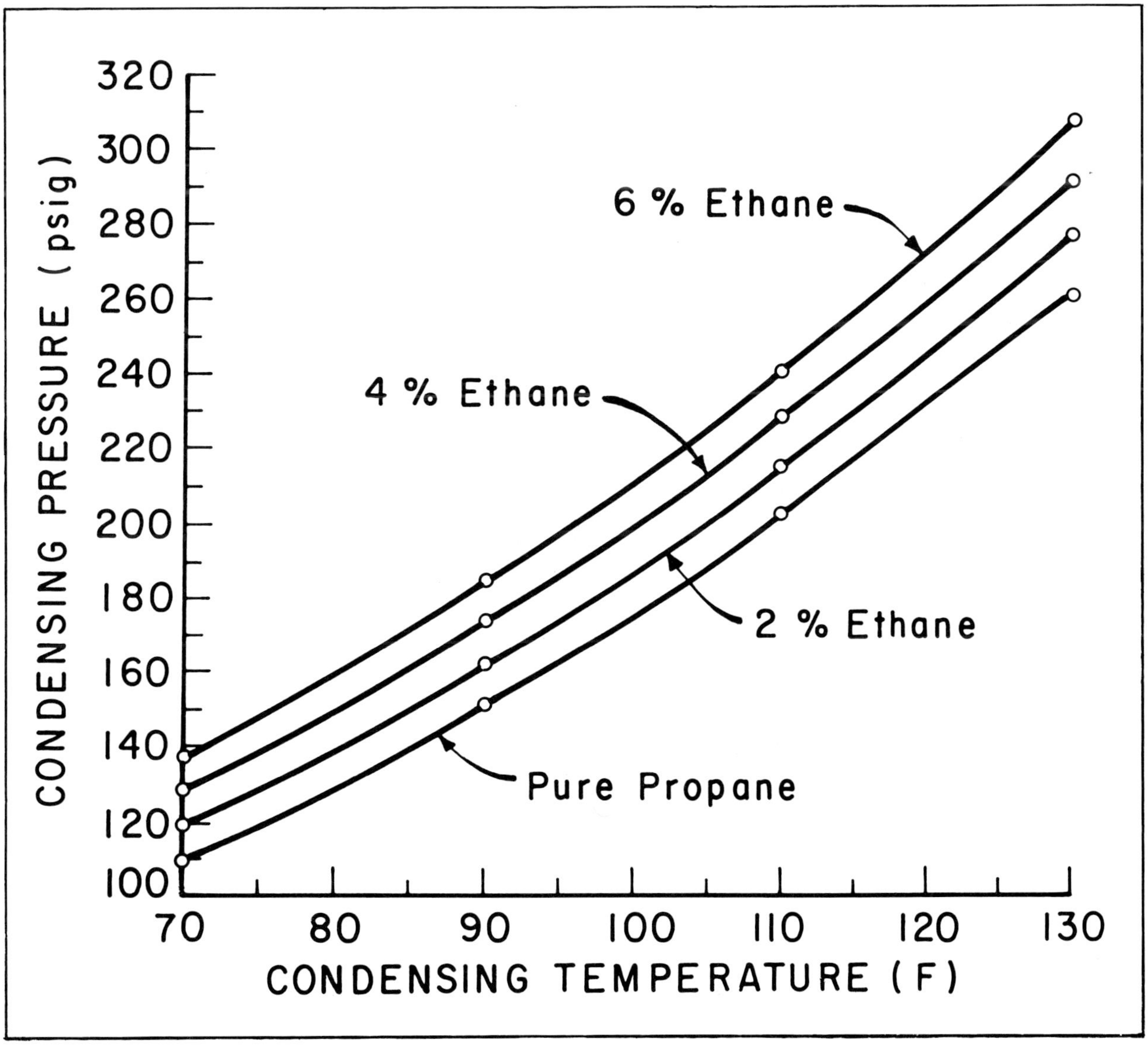

Figure 2.25. Propane condensing purity affects compressor discharge pressure.

ethane and some isobutane. Poor-quality propane impairs the performance of the refrigeration compressor. Figure 2.22, discussed previously, shows the temperature indicators (TI) and pressure indicators (PI) that must be checked so that the propane quality may be determined. These locations are at the surge tank and the chillers.

Ethane in the propane increases the compressor discharge pressure. Record the pressure and temperature of the propane surge tank and locate the point corresponding to that pressure and temperature on the graph illustrated in figure 2.25. This point should be near one of the ethane percentage curves on the graph. If the ethane content of the propane is greater than 2 percent, the excess ethane should be vented from the surge tank.

At a fixed compressor suction pressure, butane in the propane increases the chiller temperature. Figure 2.26 may be used to determine the butane content of the propane in the chillers. Record the pressure and temperature of the propane within the chiller and locate this point on the graph. Butane will be a liquid in this instance, and it should be drained from the chillers.

Efficient operation of the refrigerant condenser will achieve a minimum compressor discharge pressure. For air-cooled condensers, keep the fans, drivers, and drive assembly in good working order. For cooling-water condensers, monitoring of condenser operation is relatively simple. Maintain the cooling-water circulation to the condensers, and keep a record of two temperatures: (1) refrigerant from the condenser and (2) cooling water from the condenser. At the same time the temperatures are recorded, also note the refrigerant flow. If the refrigerant flow is not metered, record some measure of compressor load, such as the engine manifold pressure. As the cooling-water scale builds up in the condensers, these two temperatures, at the same refrigerant flow or compressor load, will be farther apart. The condensers should then be cleaned before too much product recovery is lost due to an excessive compressor discharge pressure.

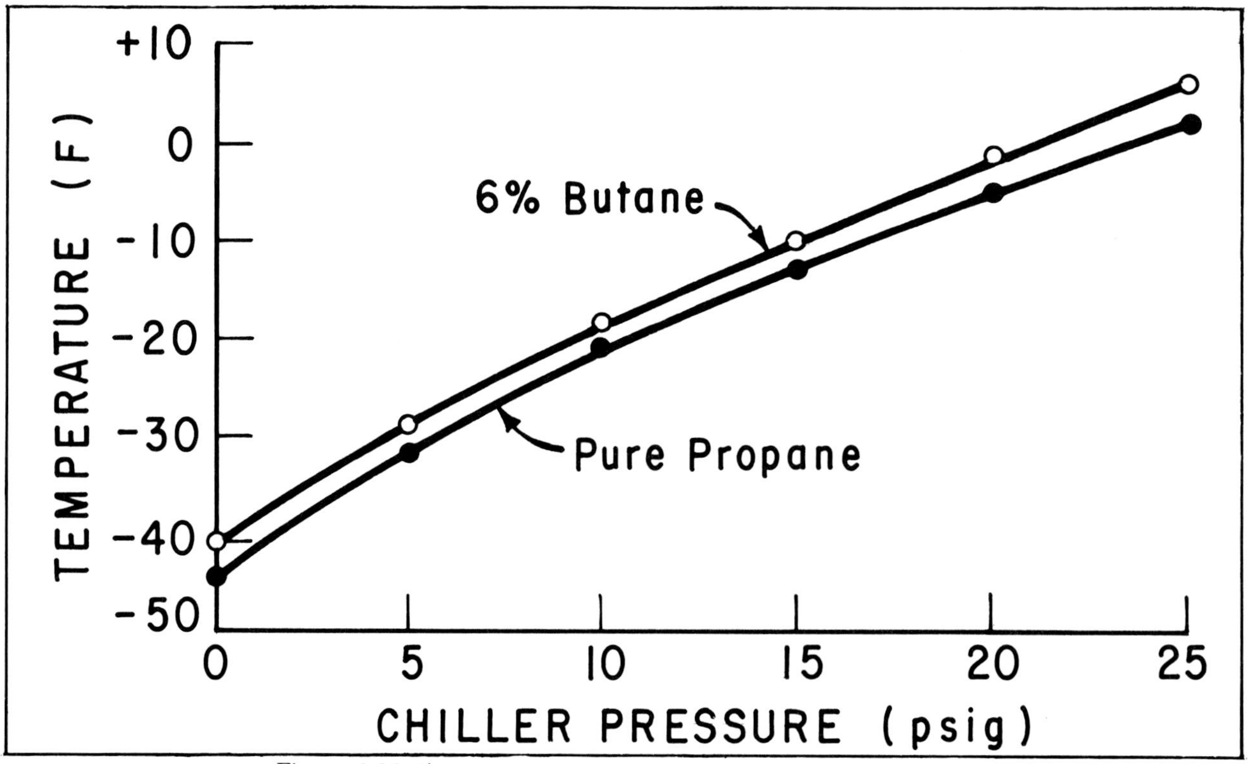

Figure 2.26. Propane chiller purity affects chiller temperature.

Chapter III

Fractionation

A system or process called fractionation is used to separate a mixture of products into individual products for sale. Fractionation is possible when the two products to be separated have different boiling points. The difficulty of fractionation can be related to the difference in the boiling points of the desired products. The usual order is to remove the lighter product first. The hydrocarbon components normally processed in a fractionation plant are shown in table 2 in the order of decreasing volatility, which is denoted by an increase in the boiling point.

Table 2. Hydrocarbon Components Processed in a Fractionation Plant

Component	*Boiling Point, F (at atmospheric pressure)*
Ethane	-127
Propane	-44
Isobutane	11
Normal Butane	31
Pentanes + Natural Gasoline	82 to 350

The essential parts of a fractionation system are the tower, condenser, reboiler, and reflux equipment. Fractionators are generally required to make relatively pure products. This means a lot of vapor and liquid contact. Getting this contact requires a large number of trays in the tower, a lot of heat input by the reboiler, and heat removal by the reflux and condenser.

Figure 3.1 shows the equipment connected into a fractionation system.

There are many methods used to control fractionators. Basically, the problem is that operating difficulties are created when both the bottom and top products are on temperature control. The reboiler temperature must be regulated to control the composition of the bottom product, and the overhead temperature (or the accumulator temperature when the top product is a vapor) must be regulated to control the composition of the top product. So, operational

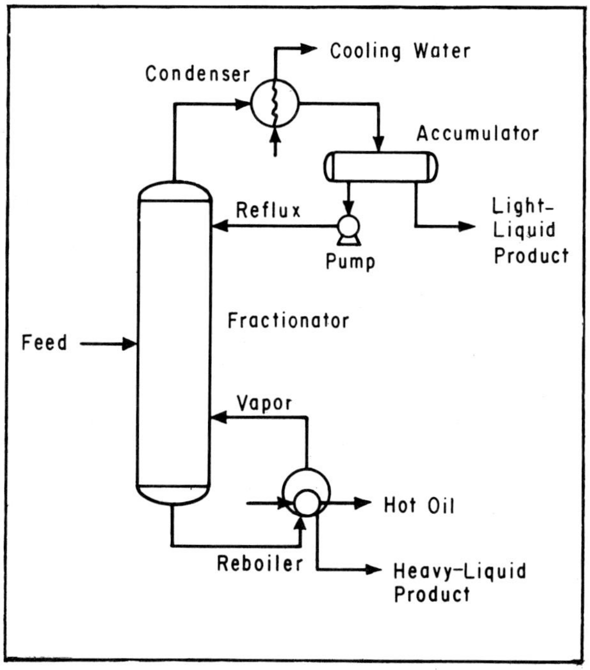

Figure 3.1. Fractionation System

control of a fractionator must be a compromise between temperature and flow regulation. Most fractionators have the reboiler heat input on bottom temperature control and the reflux on flow control. This type of control system appears to respond best to operating changes. The methods used for controlling the pressure of fractionators also differ, and these will be explained in chapter five.

FRACTIONATORS

Fractionators are usually named for the overhead, or top, product. Therefore, a deethanizer implies that the top product is ethane. A depropanizer indicates that the top product is propane. The usual fractionators follow.

Deethanizer

The first step in the fractionating sequence is to separate the ethane and propane, with the ethane going overhead and the propane and heavier components passing out the bottom of the fractionator. The operation of the deethanizer is the key to the control of the propane in the ethane and the ethane in the propane.

Depropanizer

The next step in the processing sequence is to separate the propane and isobutane, with the propane going overhead and the isobutane and heavier components passing out the bottom of the depropanizer. The operation of the depropanizer is the key to the control of the butane in the propane and the propane in the butane. In some instances, where propane and ethane are sold as a mixture, the deethanizer may be omitted, and the depropanizer may be used to separate the propane and ethane as a mixture.

Debutanizer

The next step in the sequence is to separate the normal butane and the pentanes-plus—that is, natural gasoline. The butanes (both iso and normal) pass overhead, and the pentanes-plus pass out the bottom of the fractionator. Some butane may be left in the bottom product to control the natural gasoline vapor pressure where this is required. The operation of the debutanizer is the key to the control of the pentanes in the butane and the butanes in the natural gasoline.

Butane Splitter or Deisobutanizer

When it is desirable to do so, the butanes (iso and normal), which pass overhead in the debutanizer, may be separated into iso and normal butanes. The isobutane goes overhead, and the normal butane goes out the bottom. The operation of the deisobutanizer is the key to controlling the normal butane in the isobutane and the isobutane in the normal butane.

FRACTIONATOR OPERATION

The first objective in operating a fractionator is to make products that will meet the required specifications for sale as required by the product sales contracts. It is assumed in the following discussion that there is a two-tower fractionation plant consisting of a depropanizer and a debutanizer as illustrated in figure 3.2. Each tower has an air-cooled condenser and is regulated by pressure, temperature, flow, and level controllers.

The top product of the depropanizer is propane. Propane product consists of ethane, propane, and isobutane. The bottom product is composed of a very small amount of propane, plus all of the components heavier than propane that are fed to the tower. With a compromise control system, the isobutane content of the propane product is primarily controlled by the reflux flow regulator control (FRC). An increase in reflux flow will decrease the isobutane content of the propane product. The bottom temperature regulator control (TRC) regulates the quantity of propane in the bottom product.

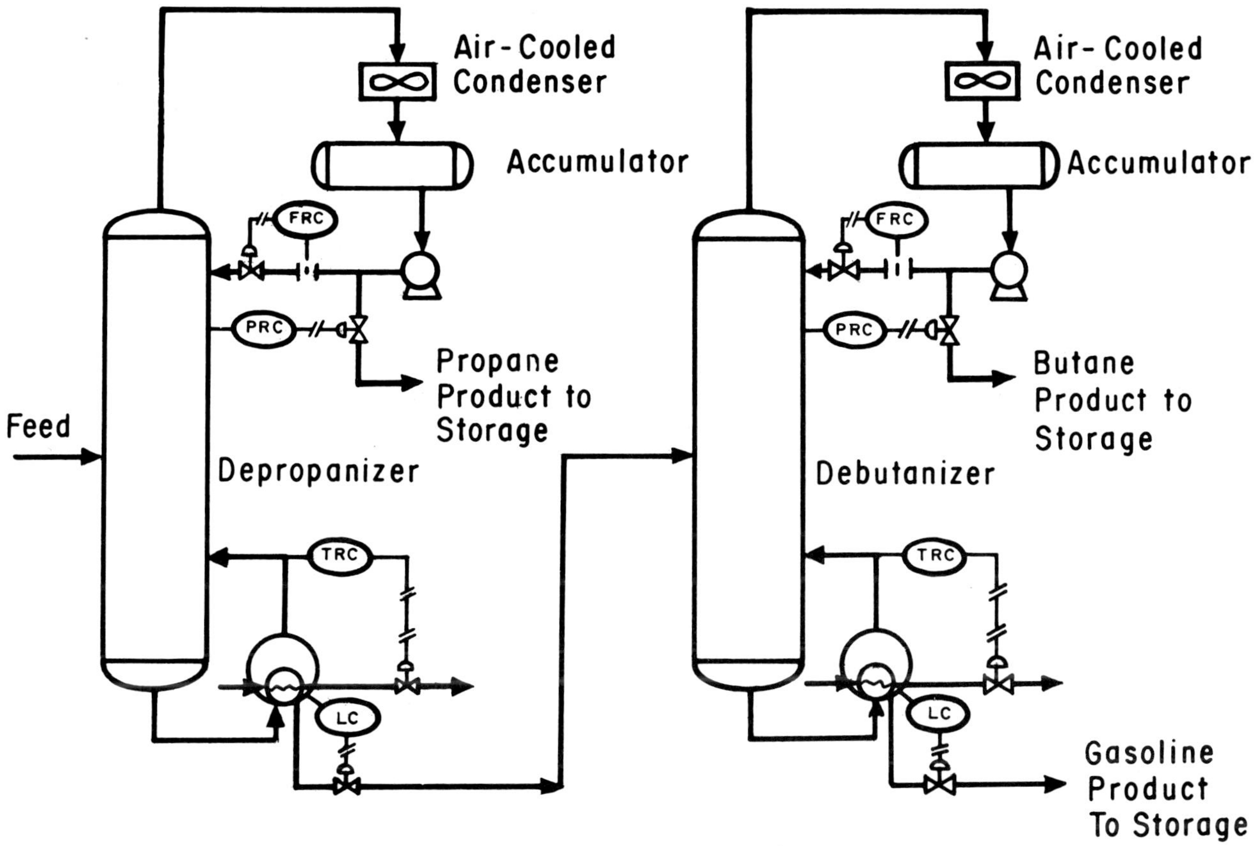

Figure 3.2. Two-Tower Fractionation Plant

Temporary tower overpressure can be caused by loss of reflux and too much reboiler heat input. Sustained overpressure can be caused by too much light component in the propane product and loss of a condenser cooling fan, which will cause reduced condenser effectiveness. The light component in the propane product is ethane. The ethane content of the propane is actually controlled upstream of the depropanizer. Therefore, adjustments in the operation of the upstream facilities are required for the reduction of ethane in the propane product.

The light component in the debutanizer overhead product is propane. Too much propane in the butane product means that the set point of the depropanizer temperature controller should be raised.

If the pressure of the depropanizer increases and the pressure cannot be quickly brought back to the pressure regulator controller (PRC) setting, then the set point on the temperature controller should be raised. If this action is not taken, then excess propane will be fed to the debutanizer and cause it to overpressure.

Tower underpressure occurs in both fractionators when too much isobutane is leaving with the propane product. Isobutane in the propane lowers the vapor pressure of the product and may cause difficulties in maintaining the depropanizer pressure. Isobutane raises the vapor pressure of the debutanizer overhead product. Therefore, when underpressure exists in both fractionators, lower the depropanizer bottom temperature. Now more isobutane will leave with the depropanizer bottoms.

POSSIBLE FRACTIONATOR PROBLEMS

The more frequent problems with fractionators are variations in feed flow, feed composition, and tower flooding.

If feed flow variations are substantial and frequent, then at low feed-flow rates, the fractionators must be operated so that very pure products are being made. This is necessary because a large increase in feed flow will cause a decrease in fractionator separation efficiency, and product purity will not be as good as at low feed rates to the fractionator.

Off-specification product will result from significant variations in feed composition, unless operating changes are quickly made. Most fractionation plants have a feed surge tank located upstream of the first fractionating tower. Often, changes in the pressure or temperature of this feed surge tank will be the first indication of a feed-composition variation.

A fractionator, like any other trayed tower, will flood when its maximum capacity is exceeded. A fractionator can be flooded by (1) too much feed; (2) too much reflux cooling from too much flow, too low a temperature, or a combination of these; and (3) too much bottom heat input. Flooding destroys separation efficiency. When flooding occurs, reflux will be carried overhead as a liquid. If good purity products were being made prior to flooding, then reduce the tower load by decreasing the reflux cooling. If, prior to flooding, product purity was just sufficient to meet sales specifications, then reduce the tower load by decreasing the feed flow.

Chapter IV

Dehydration

All hydrocarbon fluids can retain some water. Water is soluble in liquid hydrocarbons and can be held in the vapor phase by hydrocarbon gases. As the liquid or gas is cooled, it cannot retain water as well. Figure 4.1 shows how the water content of natural gas varies with temperature and pressure. For instance, at 1,000 psia and 100 F, natural gas holds about 62 lb of water per million cubic feet of gas. As the gas is cooled, its ability to retain water decreases. At 1,000 psia and 0 F, the gas contains only about 2 lb of water per MMcf of gas.

Note that the title in the lower right-hand corner of figure 4.1 says, "dew point of natural gas." This title is a little misleading in that the chart actually applies to the dew point of the water content in natural gas. The water dew point of a gas (or liquid) is that temperature at which free water will begin to separate from the gas (or liquid). In the example, if a natural gas stream at 1,000 psia contains 62 lb of water per MMcf of gas, its water dew point is 100 F. If it is cooled below 100 F, then free water will be present.

Hydrates will form if a gas or liquid containing free water is cooled below its hydrate temperature. Figure 4.2 can be used to estimate the hydrate temperature of natural gas. For instance, a 0.6 specific gravity gas has a hydrate temperature of about 61 F at 1,000 psia. If this gas is to be cooled below its hydrate temperature, either due to pipeline transportation, pressure reduction for consumption, or for processing, then provisions must be made for avoiding free-water problems that cause freezing and the formation of hydrates.

Hydrates are solid compounds that form as crystals and resemble snow in appearance. They are created by a reaction of natural gas with water, and when formed, they are about 10 percent hydrocarbon and 90 percent water. Hydrates have a specific gravity of about 0.98, and they will usually float in water and sink in hydrocarbon liquids. Water is always necessary for hydrates to form.

Freezing can be avoided by either removing the water from the gas or liquid prior to cooling below the hydrate temperature or by using a hydrate inhibitor to mix with the water condensed during cooling. Dehydration is the removal of water from a substance. Dehydration may be accomplished by the use of solid substances such as those used in dry-bed dehydrators; or a liquid, such as triethylene glycol, may be employed; or a gas may be used, such as the stripping gas in a triethylene glycol (TEG) reboiler. The more common hydrate inhibitors are methanol and ethylene glycol. Most natural gas facilities use only one or more of the following: glycol (ethylene glycol) injection, TEG (triethylene glycol) dehydration, or dry-bed dehydrators.

GLYCOL INJECTION

Figure 4.3 depicts a typical glycol (EG) injection system for a refrigerated gas plant. Lean—that is, concentrated—glycol is injected into the inlet gas upstream of both the gas/gas exchanger and gas chiller. Figure 4.4 shows ethylene glycol injection directly into the gas stream. The glycol mixes with the water that is condensed as the in-

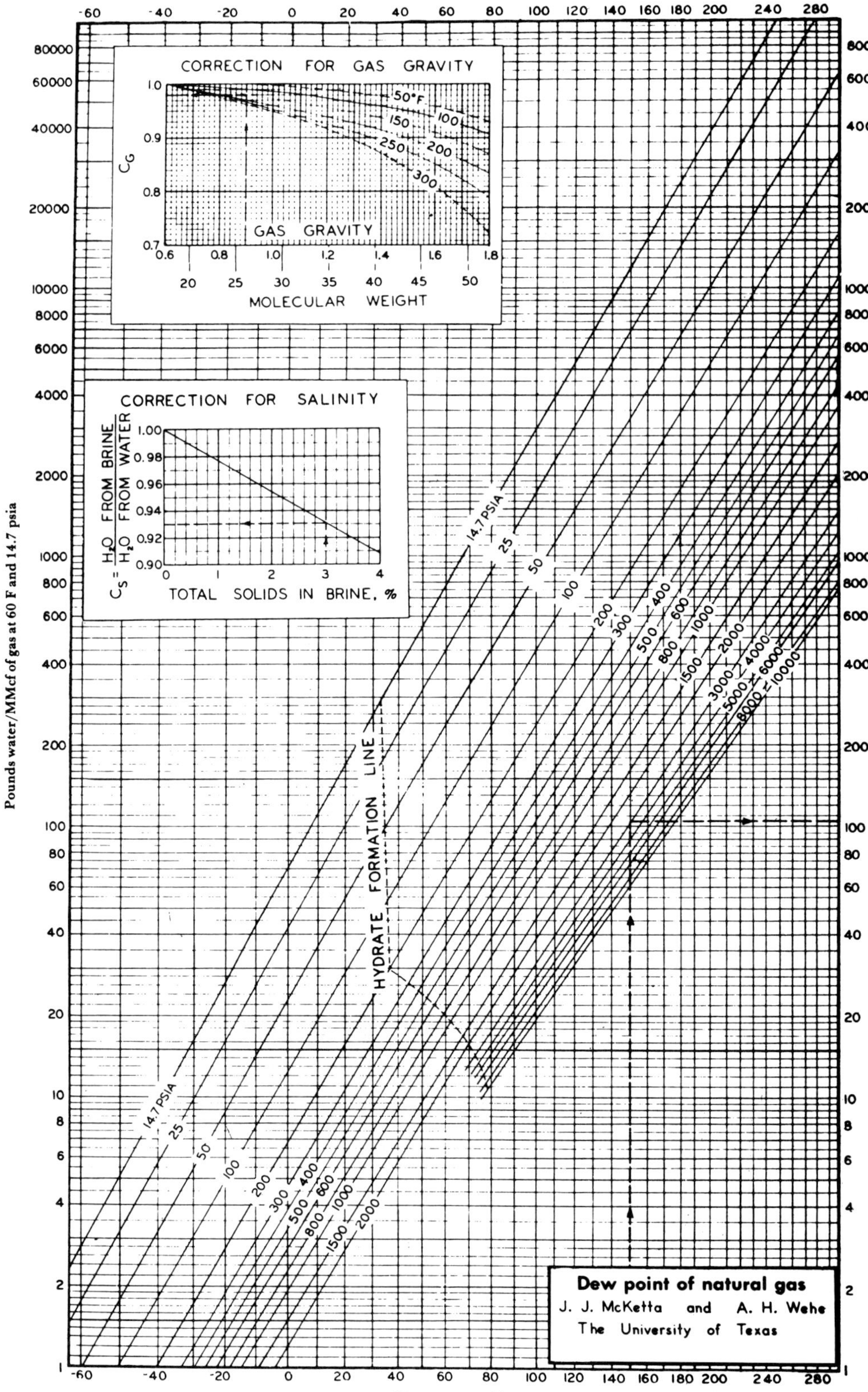

Figure 4.1. Water content of natural gas varies with pressure and temperature.

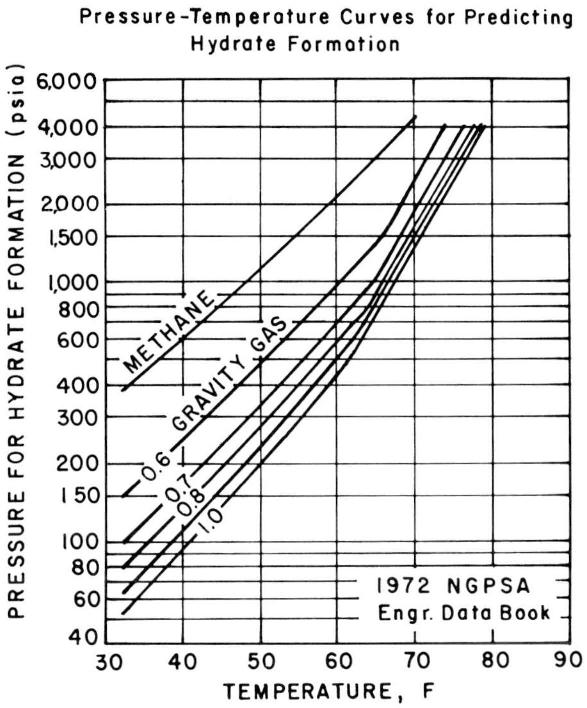

Figure 4.2. Hydrate Temperatures at Various Pressures for Different Gases

Figure 4.4. Arrows show glycol injection points.

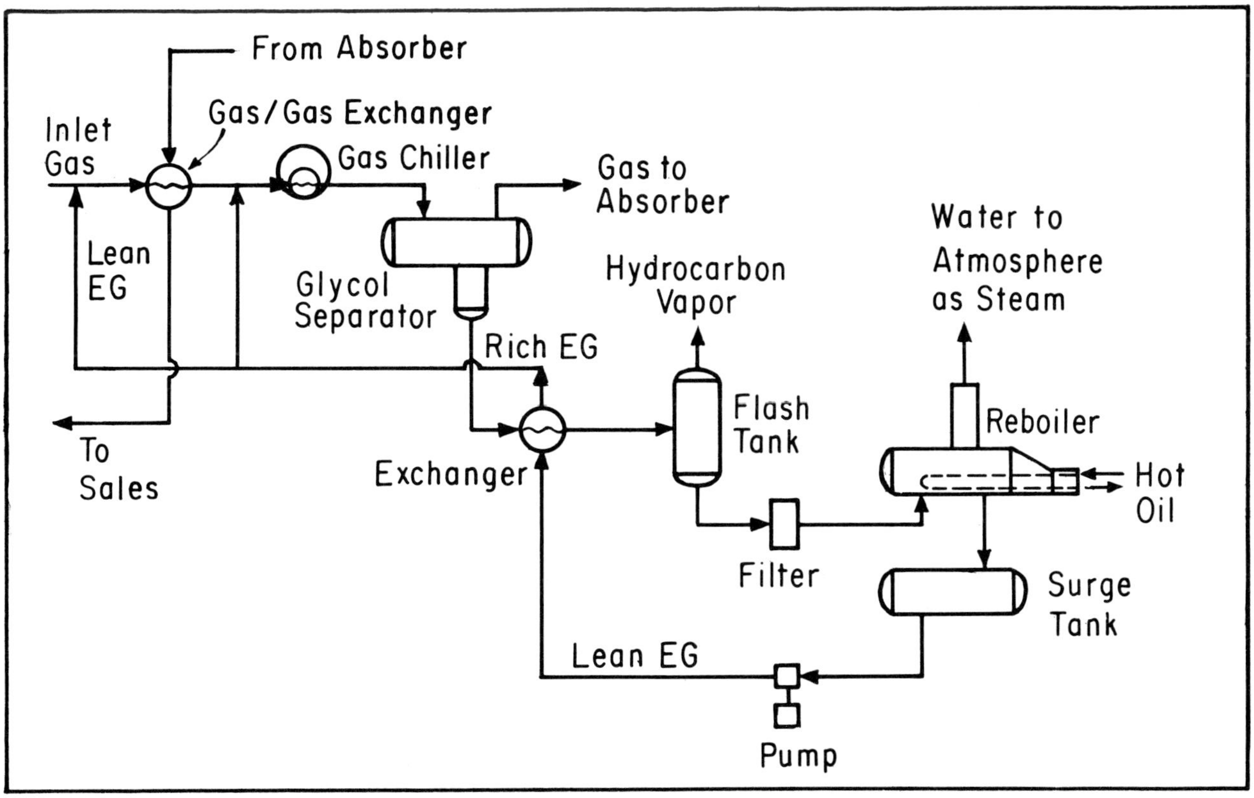

Figure 4.3. Ethylene Glycol (EG) Injection Dehydration Unit

let gas is cooled. The glycol prevents freezing in the same way that antifreeze prevents the water system in a car engine from freezing during cold weather. The glycol-water mixture, called *rich glycol,* is separated from the hydrocarbon vapor and liquid in the glycol separator.

The glycol is warmed in the glycol exchanger and fed to the flash tank. The flash tank is normally operated at about 60 psig. This combination of heat addition and pressure reduction between the glycol separator and flash tank rejects the light hydrocarbons (primarily methane) absorbed by the glycol.

The glycol is regenerated (concentrated) by vaporizing the extra water in the reboiler. A glycol regenerator is shown in figure 4.5. Heat is supplied to the reboiler by a fire tube that burns fuel gas, a hot circulating fluid such as hot oil, or by steam. The lean glycol is then pumped from the surge tank, through the glycol exchanger, and again injected into the inlet gas.

Figure 4.5. Glycol Regenerator

The majority of glycol injection unit problems can be divided into three categories:

1. glycol concentration;
2. glycol circulation; and
3. glycol solution condition.

Highly concentrated glycol is not desirable in an ethylene glycol injection system. In fact, figure 4.7 shows that 100 percent glycol can cause freezing problems. The lean glycol concentration should be a maximum of about 80 percent. This 80 percent concentration gives good freeze protection and maintains reasonable glycol

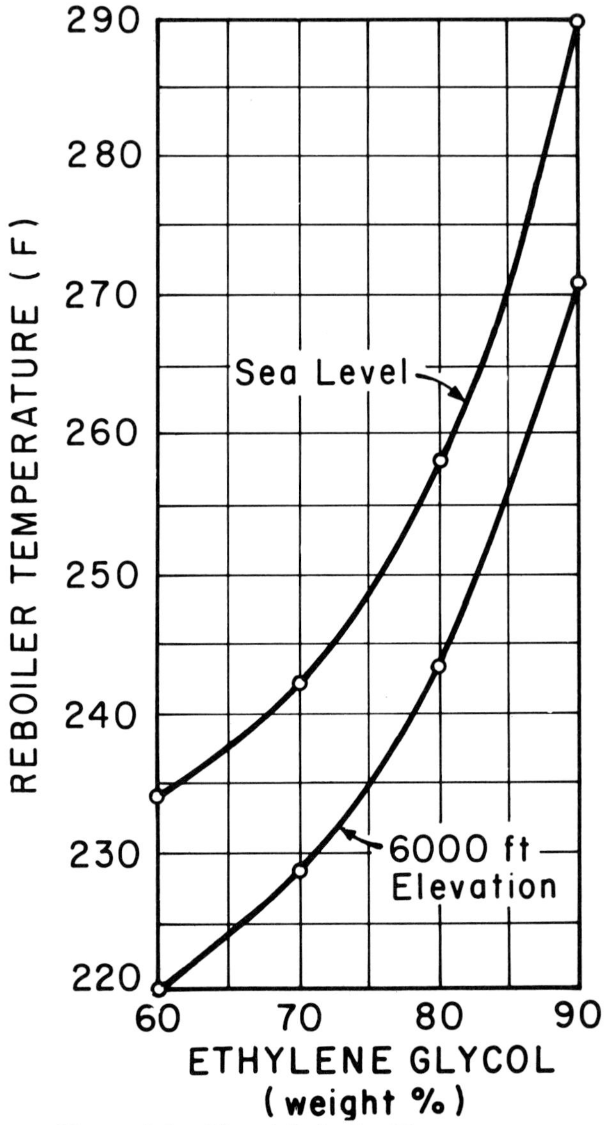

Figure 4.6. Glycol Reboiler Temperatures

losses from the reboiler. The proper reboiler temperature depends on plant elevation. The correct reboiler temperature for the desired glycol concentration may be read from figure 4.6.

The glycol circulation rate must be adequate for good distribution of the glycol in the inlet gas and to maintain the rich glycol concentration. The concentration of rich glycol from the glycol separator depends on the lean glycol concentration, the glycol circulation rate, and the quantity of condensed water. The rich glycol concentration must be sufficient to prevent freezing. Figure 4.7 can be used to determine the freezing temperature of the rich glycol solution. However, the *hydrate depression ability* of the rich glycol must also be examined.

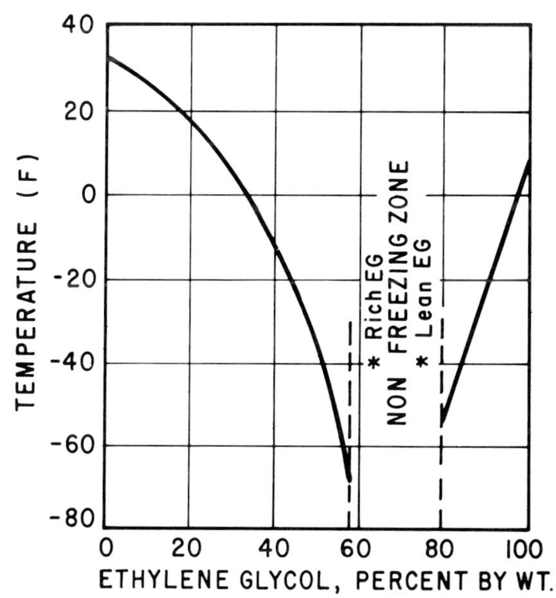

Figure 4.7. Freezing Temperatures of Ethylene Glycol-Water Mixtures

Assume a 0.6 specific gravity gas at 1,000 psia is to be cooled to 0 F. This gas has a hydrate temperature of 61 F, as can be determined from the Gas Hydrate Temperature Graph shown previously in figure 4.2. The hydrate depression required is 61 minus 0 or 61 F. Using the graph in figure 4.8, it can be seen that a 61 F hydrate depression requires a 62 percent glycol concentration. This means that sufficient lean glycol must be circulated to maintain a minimum rich glycol concentration of 62 percent.

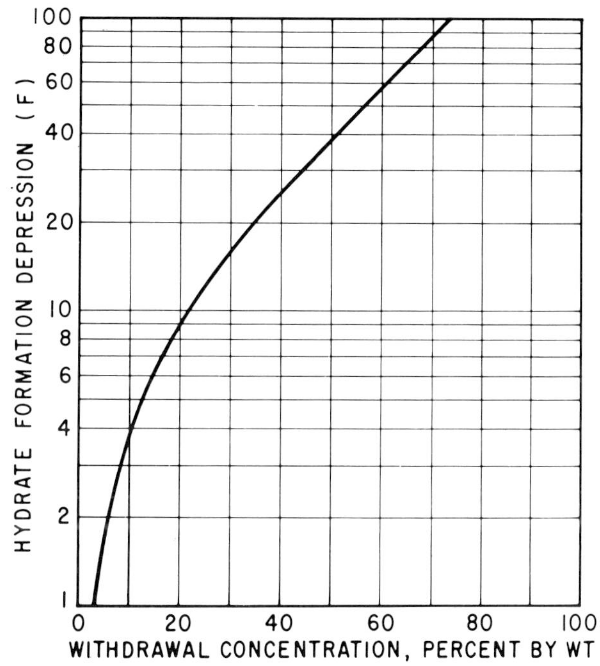

The Dow Chemical Co. – Gas Conditioning Fact Book

Figure 4.8. Hydrate Depression versus Minimum Withdrawal Concentration of Ethylene Glycol

Glycol solution condition depends on the quantity of solid material and heavy hydrocarbons. Solid material must be continuously filtered from the circulating glycol stream. Do not extend the time between filter-element replacements by periodically bypassing the filter. If frequent filter-element replacements are required, use a large-mesh element for two or three filter changes, and then use a small-mesh element for one filter change. The use of a small-mesh element for one out of every three or four filter changes normally controls the buildup of solid material in the circulating glycol stream.

Poor separation of the rich glycol from the hydrocarbon liquid in the glycol separator will increase glycol losses and cause contamination of the glycol. Hydrocarbons in the glycol can cause foaming problems. The separation efficiency in the glycol separator primarily depends on residence time and temperature. Longer residence time and higher temperatures improve separation. If separator efficiency cannot be improved to a satisfactory level, the installation of a charcoal filter on the glycol-pump discharge adequately maintains glycol solution condition.

TRIETHYLENE GLYCOL (TEG)

A TEG dehydration system removes the water from the gas (or liquid) by absorption. Figure 4.9 illustrates a TEG dehydration system. Generally, all of the equipment, except the contactor, is package- or skid-mounted.

The contactor is really an absorber tower. Lean TEG, which has a small water content, absorbs water from the inlet gas and leaves the bottom of the contactor rich with water. Since the TEG also absorbs some of the light hydrocarbon vapor, primarily methane, the pressure of the rich TEG is reduced through a flash tank to allow most of these hydrocarbon vapors to escape. The TEG is then filtered to remove solids, heated in an exchanger, and fed to the shell of the reboiler. The reboiler may be hot-oil heated, steam heated, or direct fired. The TEG is reconcentrated by removal of the water in the reboiler. The smaller the water content of the lean TEG, the lower the water content of the dehydrated gas leaving the contactor.

If it becomes difficult to adequately dehydrate the inlet gas, the following operating items need to be checked:

1. glycol concentration;
2. glycol circulation; and
3. glycol solution condition.

The glycol concentration required for proper dehydration can be checked two ways. Assume that the dehydrated gas is to have a maximum water content of 5 lb of water per MMcf of gas when delivered at a pressure of 985 psig (1,000 psia). By referring back to figure 4.1, it can be

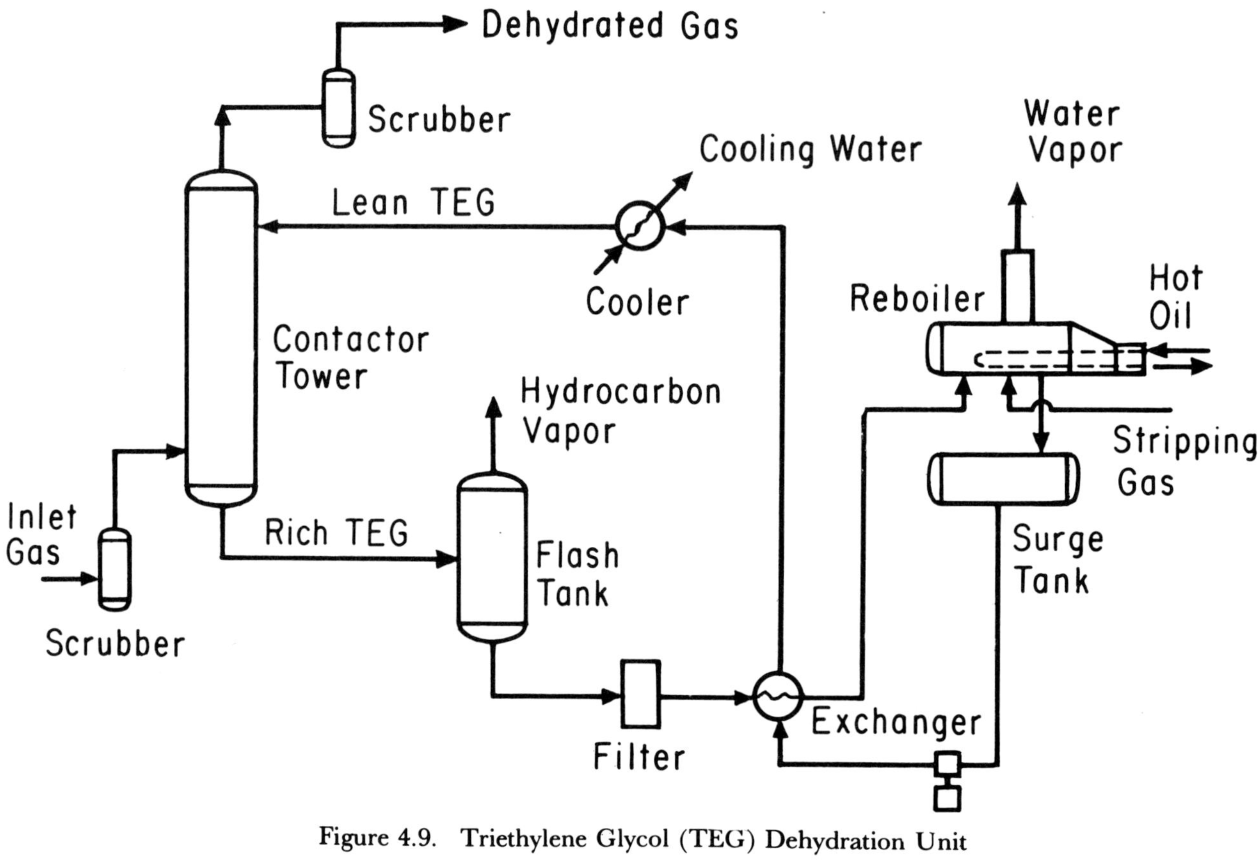

Figure 4.9. Triethylene Glycol (TEG) Dehydration Unit

seen that 5 lb per MMcf and the 1,000 psia line intersect at a water dew point temperature of 23 F. Assume that a dehydrated gas leaves the contactor at 100 F. From figure 4.10 it can be seen that a 23 F dew point at 100 F contact temperature requires a TEG concentration of about

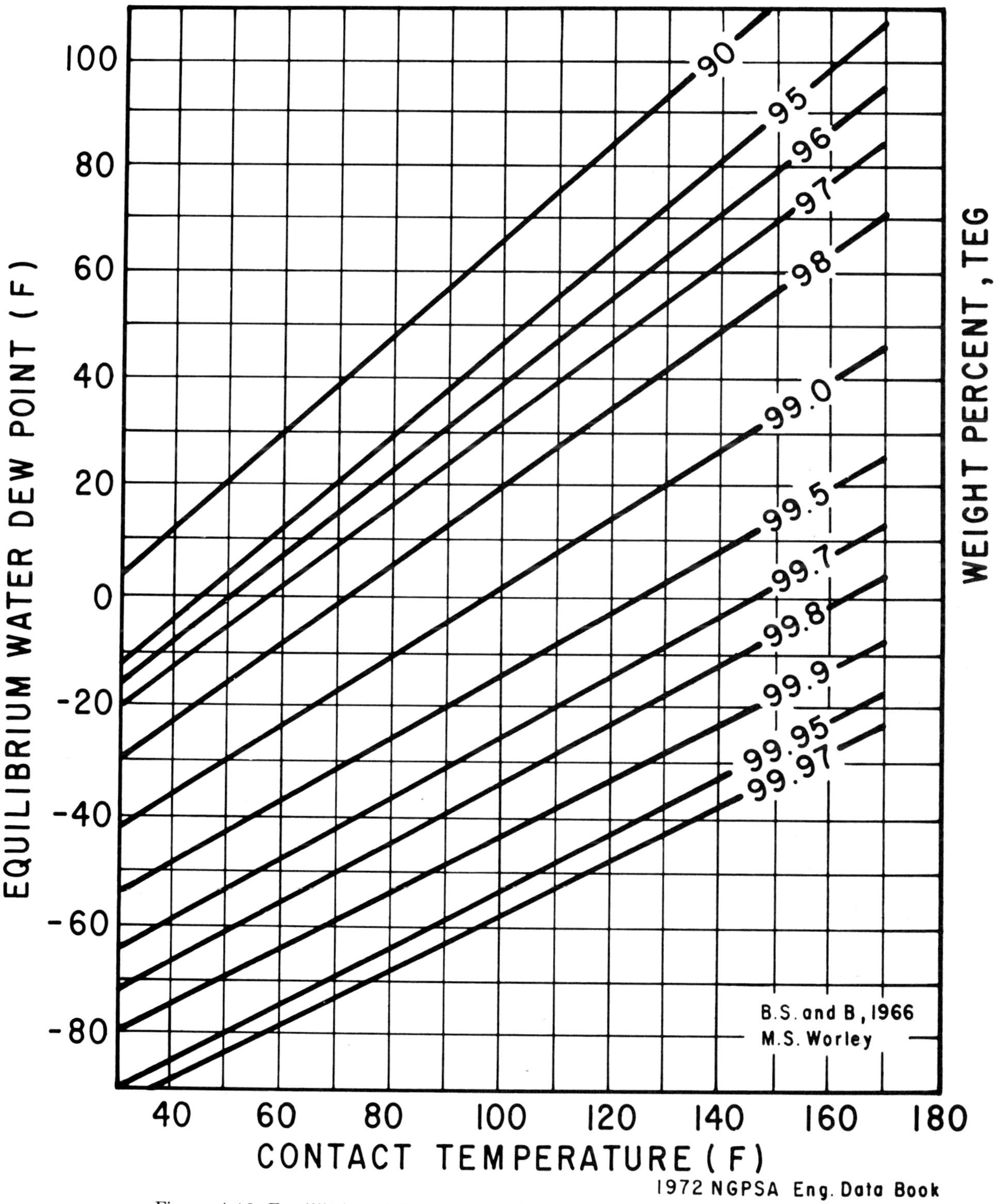

Figure 4.10. Equilibrium Water Dew Points with Various Concentrations of TEG

97.7 percent. Next, looking at figure 4.11, it can be determined that a 97.7 percent TEG concentration requires a solution temperature in the reboiler of 367 F, at sea-level elevation.

If the TEG unit is upstream from a refrigerated gas processing plant, then the procedure for determining the proper glycol concentration is slightly different. Use the refrigerant temperature in the gas chiller as the required dew point temperature. The main gas stream does not get chilled to the refrigerant temperature, but part of the gas, specifically that part next to the cold tubes, will get close to the refrigerant temperature. Assume the refrigerant temperature is -10 F. Figure 4.10 shows that a -10 F dew point at a contact temperature of 100 F requires a TEG concentration of about 99.4 percent. The required reboiler temperature can now be read from figure 4.11. However, figure 4.11 shows that the required reboiler temperature is above 440 F at sea level, and manufacturers of TEG recommend a maximum TEG temperature of 400 F. Above 400 F, the TEG will begin to decompose or degrade—that is, break down into other components. So, the maximum reboiler temperature should be 400 F. The 99.4 percent TEG concentration can be attained by either operating the reboiler at a vacuum—the lower pressure lowers the required reboiler temperature just as operating at higher elevations does—or by using a dry, low-pressure gas, called *stripping gas.* Most TEG units requiring high glycol concentrations use stripping gas.

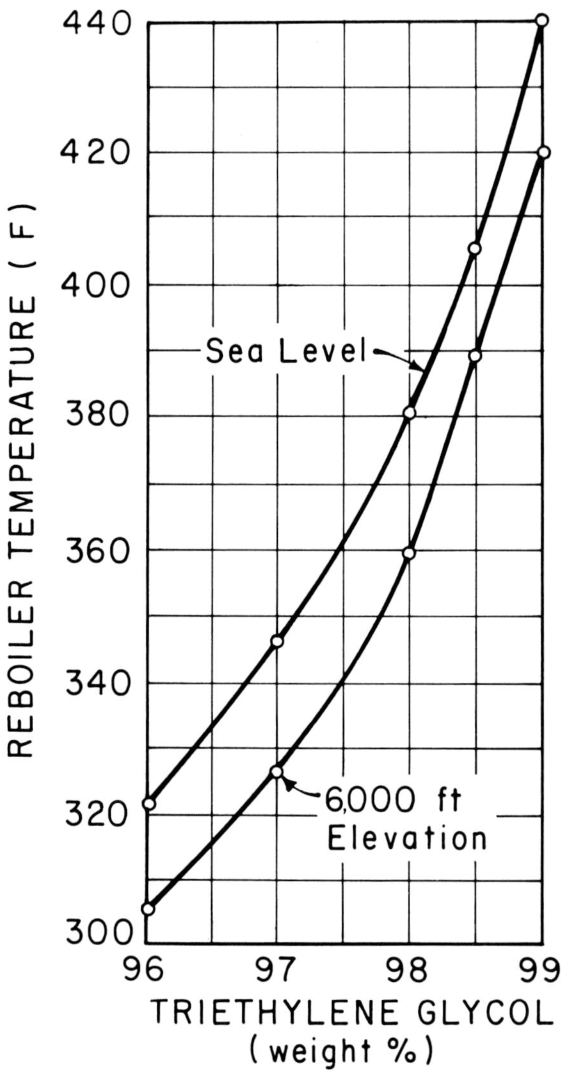

Figure 4.11. Triethylene Glycol Reboiler Temperatures

The reboiler, when operated at 400 F, will vaporize and drive off most of the water. Stripping gas is used for final water removal. Figure 4.12 shows the effect of stripping gas on TEG concentration. However, it should be remembered that excess stripping gas will increase TEG losses and waste gas.

TEG circulation is important because there is a minimum withdrawal concentration necessary for proper dehydration. In the previous examples, the lean TEG concentration required was used as the proper reboiler concentration. However, the water dew point of the dehydrated gas is actually determined by the lean TEG concentration leaving the top tray of the contactor. Therefore, the concentration of the rich TEG from the contactor must be sufficiently high to prevent excess water from reaching the top tray of the contactor and diluting the effectiveness of the lean TEG.

Most of the reboiler's heat input is required for vaporizing the water. Therefore, increasing the TEG circulation will not require a large amount of extra heat. The two charts in figure

4.13 show that there is a reasonable limit to the effectiveness of increasing circulation.

The comments on glycol solution condition under the section on glycol injection also apply to TEG units. In addition, there are two other possibilities of problems with solution condition. TEG can cause condensation of some heavier hydrocarbons in the contactor. These heavy hydrocarbons are often difficult to remove from the TEG solution. Heavy hydrocarbon problems can be minimized by keeping the temperature of the TEG to the contactor about 10 F to 20 F above the temperature of the inlet gas to the contactor.

Most TEG units operate at reboiler temperatures close to the decomposition temperature. Decomposed TEG will have a pH of about 3 to 4. Normally, decomposed TEG can be seen in the gauge glass of the surge tank. A black layer floating on top of the TEG indicates heavy hydrocarbons. A black layer under the TEG indicates decomposed glycol.

DRY-BED DEHYDRATORS

Dry-bed dehydration is an *adsorption process*—that is, it uses a solid to remove water from the fluid. A sponge is a good example of adsorption. If water is spilled on the floor, a sponge can be placed in the water, and it will soak up (adsorb) the water. However, if a lot of water has been spilled and only a small sponge is available, it will be found that the sponge can only adsorb so much water before it becomes saturated. To clean up all the spilled water, the sponge must be saturated, the water squeezed out, and the sponge saturated again. In squeezing the water from the sponge, the sponge is being regenerated because it is being made to again have the capacity to adsorb water.

A dry-bed dehydration unit works similarly to the sponge. The solid material used to adsorb the water is called a *desiccant.* Typical desiccants are silica gel, Sorbead, activated alumina, and molecular sieves. When the desiccant

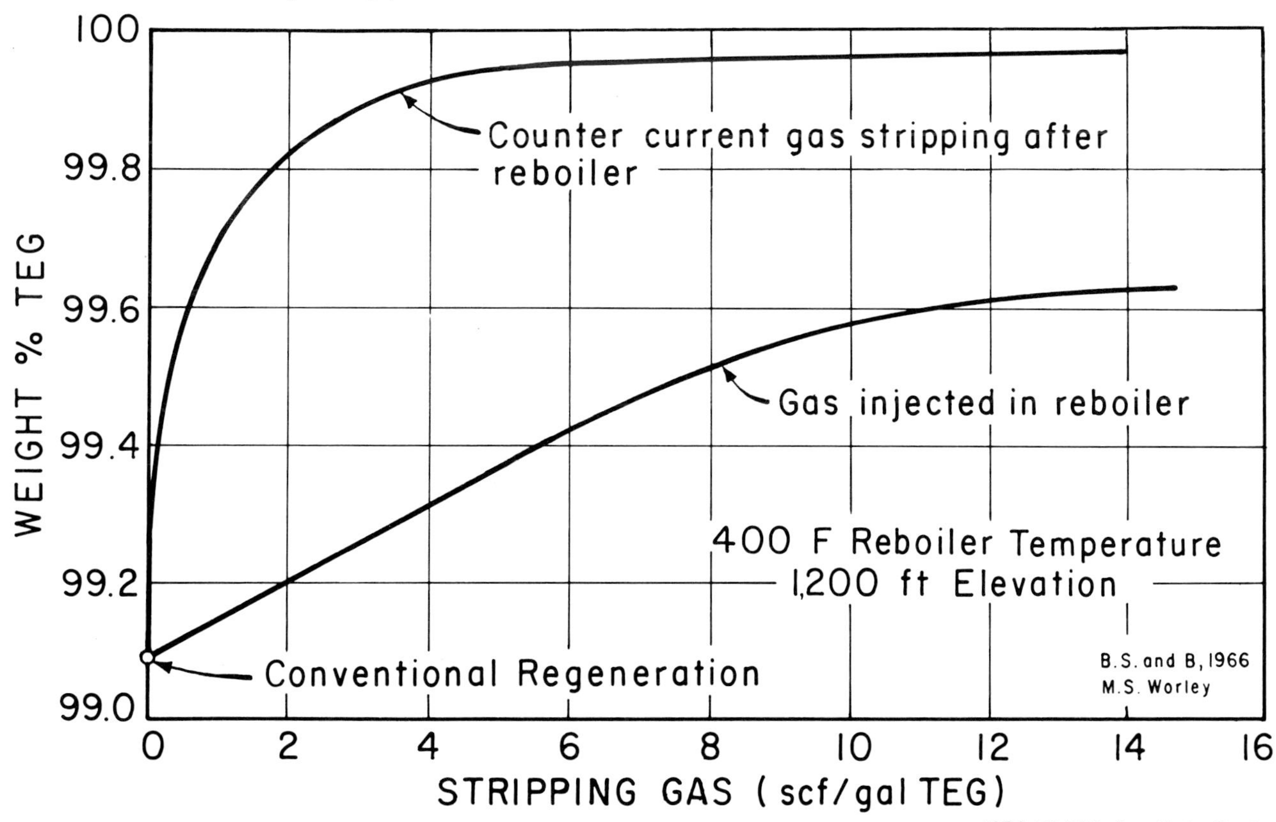

Figure 4.12. Effect of Stripping Gas on TEG Concentration

has become saturated, it must be regenerated to restore its adsorptive capacity. Regeneration of a desiccant is usually accomplished by heating. Hot gas vaporizes the water from the desiccant. A dry-bed dehydrator generally has at least two vessels filled with desiccant. One bed is adsorbing while the other bed is regenerating. In this way, gas dehydration is a continuous process.

Figure 4.14 is a flow diagram for a two-tower, dry-bed dehydration unit. The solid lines represent the gas being dried, while the dashed lines represent the regeneration gas flow. One bed is on the adsorption cycle, while the other bed is being regenerated. Regeneration requires both heating and cooling. After the desiccant has been heated to drive off the water, the regeneration gas heater, which may be a fired heater or a shell-and-tube heat exchanger using steam or hot oil, is bypassed, and the bed is cooled before it is switched to adsorption. Adsorption flow is usually down (that is, from top to bottom) to pass through the desiccant bed. Heating and cooling flow may be either up or down. Usually, regeneration flow is countercurrent to adsorption flow.

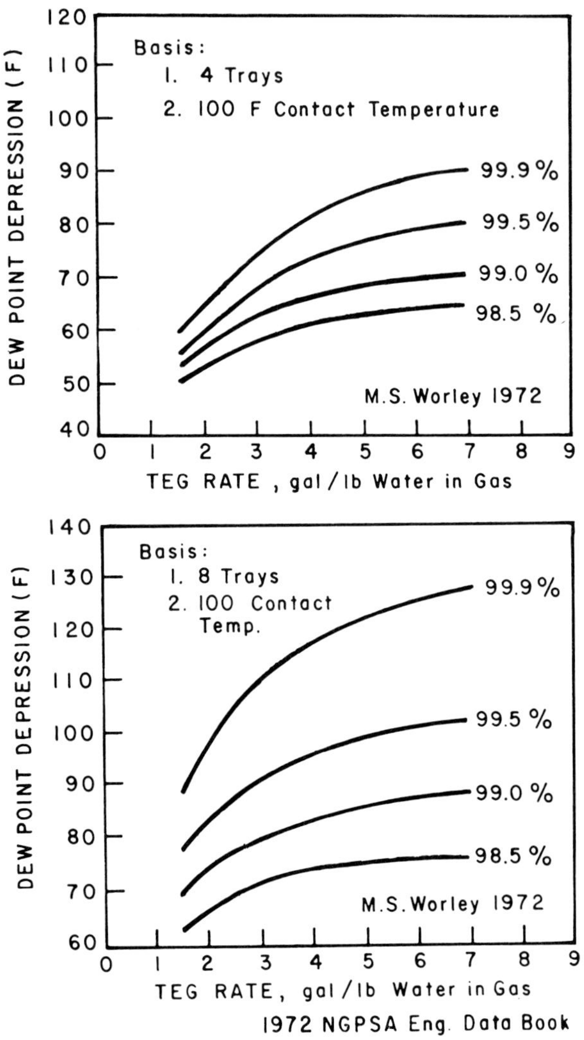

Figure 4.13. Dew Point Depression versus TEG Circulation Rate

A bed must be properly regenerated, or its adsorptive capacity is reduced. If each time the water is squeezed from the sponge, it is only "half-squeezed," then the sponge cannot hold very much additional water. If each time a bed is regenerated, it is only "half-regenerated," then it cannot adsorb as much additional water. Proper regeneration depends on both the quantity and temperature of the regeneration gas. The higher the regeneration temperature, the smaller the required quantity of regeneration gas. However, if the temperature is too high (above 600 F for silica gel, Sorbead, and activated alumina), then the desiccant will be ruined.

When dehydration becomes inadequate to meet downstream requirements, an operator must either shorten the adsorption cycle or increase the regeneration gas flow or do both. These changes should allow continued operation while the cause of the deficiency is being determined.

A three-tower, dry-bed dehydration upstream of a cryogenic gas processing plant is shown in figure 4.15. A three-tower unit may be a combination of either (1) one bed on adsorption, one bed on heating, and one bed on cooling or (2) two beds on adsorption and one bed on regeneration.

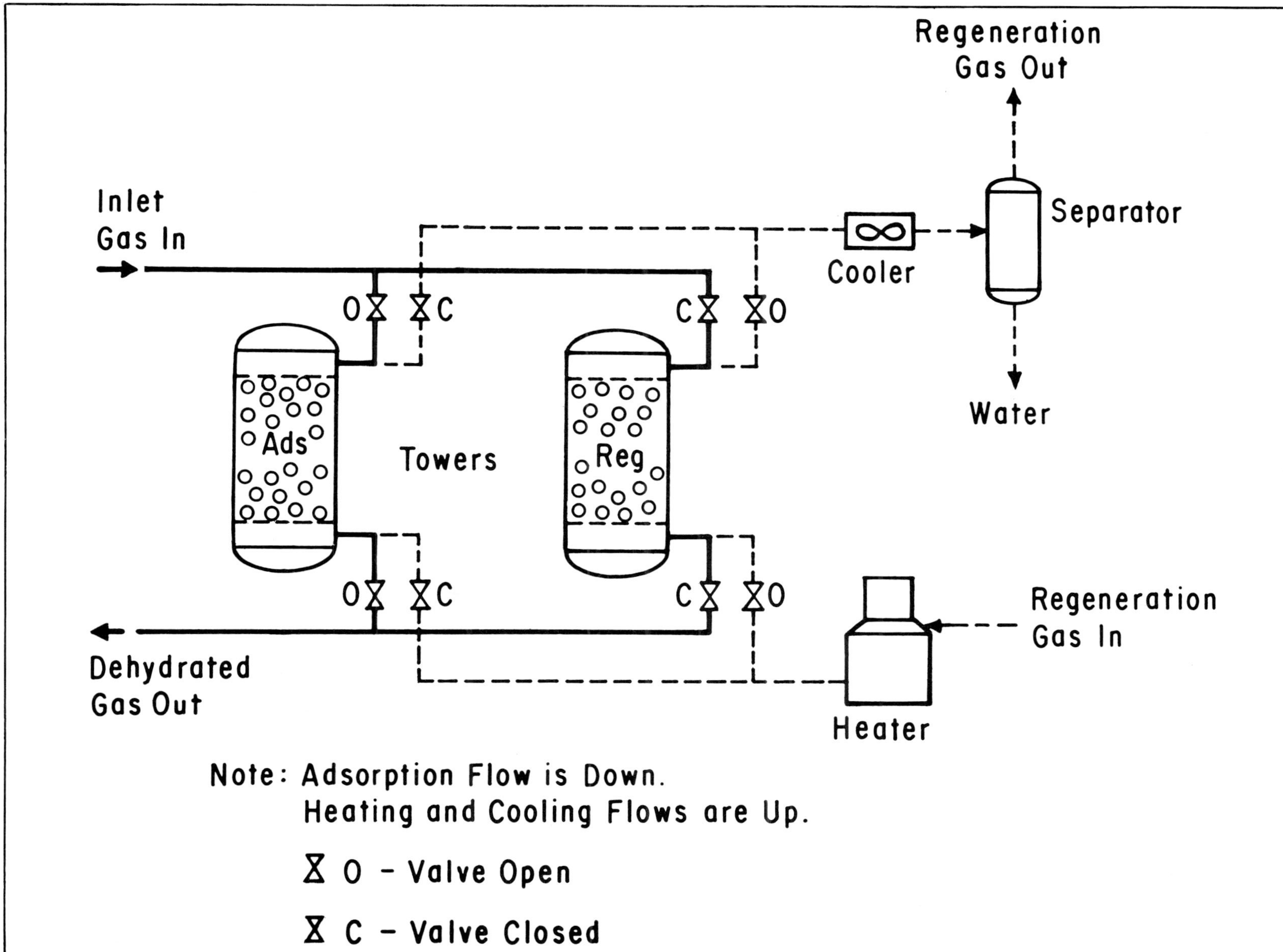

Figure 4.14. Dry Bed Dehydration Unit

Figure 4.15. Dry Bed Dehydrator

Chapter V

Instrumentation

All functions described in this chapter are related to the control of a gas or liquid; these two items are referred to as the controlled medium. The end result is to maintain the controlled medium at a specific pressure, temperature, flow rate, or liquid level. These last items then are referred to as controlled variables. Final control is accomplished either directly or indirectly by increasing or decreasing the size of the opening through which the controlled medium or some allied agent is passing. Most protective devices for gas compressors and other field equipment operate on these same principles.

MEASUREMENT OF CONTROLLED VARIABLES

Before any variable can be controlled, its value must be determined and compared to an optimum value in order to take any needed corrective action. Corrective action may be made manually by operating personnel after visually checking the value of the variable as indicated by a measuring instrument or an electrical or mechanical device. The operating personnel compare the measured value with an optimum value and automatically make the required correction.

Pressure Measurements

Although pressure can be any force from less than atmospheric (vacuum) to thousands of pounds above atmospheric, the discussion will be limited to a measurement above atmospheric. However, the instruments described can also be used to measure vacuum.

The *manometer* is the most commonly used instrument for measuring pressures from zero through a few pounds per square inch. This device is a U-shaped tube made of glass or other transparent material as illustrated in figure 5.1. The inside diameter of the tube is unimportant provided the size is consistent over its entire length. A scale with suitable markings is attached behind the unit, and the tube is filled approximately half full of fluid. Each scale is marked for a fluid having a particular specific gravity, and a fluid of any other gravity will give an incorrect reading. One end of the manometer is connected to the pressure to be measured, and the other end is open to at-

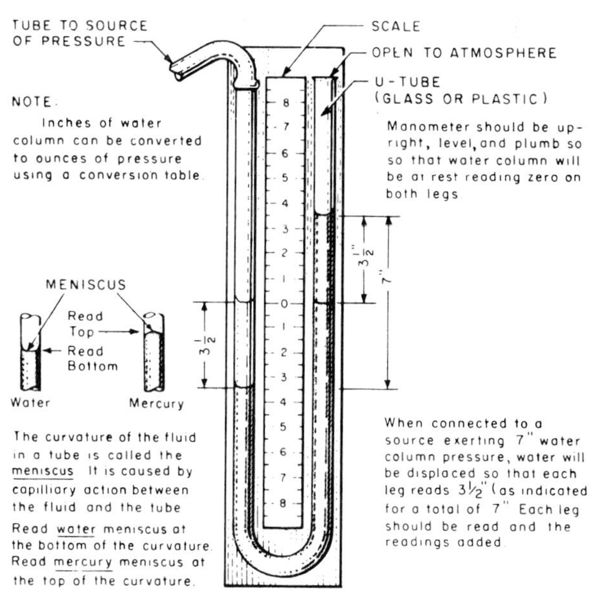

Figure 5.1. Water Manometer

mosphere. The pressure will force the fluid to move in the tube until the weight of the displaced fluid is equal to the force exerted by the pressure. The difference between the height of the fluid in the two sides of the manometer is the pressure measurement. Scales are generally marked in millimeters, inches, or ounces. Manometers are used almost exclusively in measuring pressure where visual readings are required and low pressure is involved.

A *diaphragm element* is also used in measuring low pressures up to a few pounds. It is extremely sensitive to small pressure changes and can provide an indicated reading or the motion required to operate an automatic control device to be discussed later. The basic diaphragm unit is shown in figure 5.2 and consists of a heavy metal base, a pressure connection, and a corrugated diaphragm. The diaphragm is generally

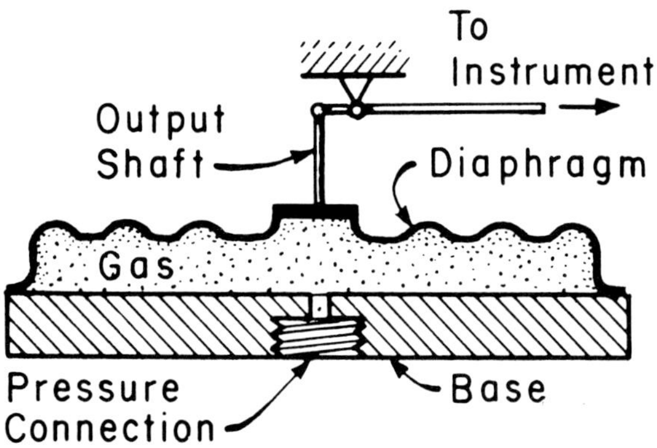

Figure 5.2. Single-Diaphragm Pressure-sensing Element

made of thin sheet metal formed to shape by a hydraulic press. However, nonmetallic materials are available to measure extremely low pressures. Corrugations allow considerably more movement than could be expected from a flat sheet for any applied pressure. Pressure applied inside the unit expands the diaphragm, and this motion is transmitted through suitable linkage to an indicating pointer, a recording pen, or the detection mechanism of an automatic controller. A special type of diaphragm element is used for applications such as aneroid barometers. If the pressure inlet is sealed, any variation in atmospheric pressure acting on the outside of the diaphragm will cause a pointer movement proportional to the change in atmospheric pressure. The motion from a single-diaphragm unit is limited, and the process of magnifying this motion with additional linkage can introduce errors and create problems of calibration. The additional movement needed to position a pointer along a longer indicating scale can be achieved by adding additional diaphragms in series to obtain an extremely sensitive and accurate measuring device (fig. 5.3).

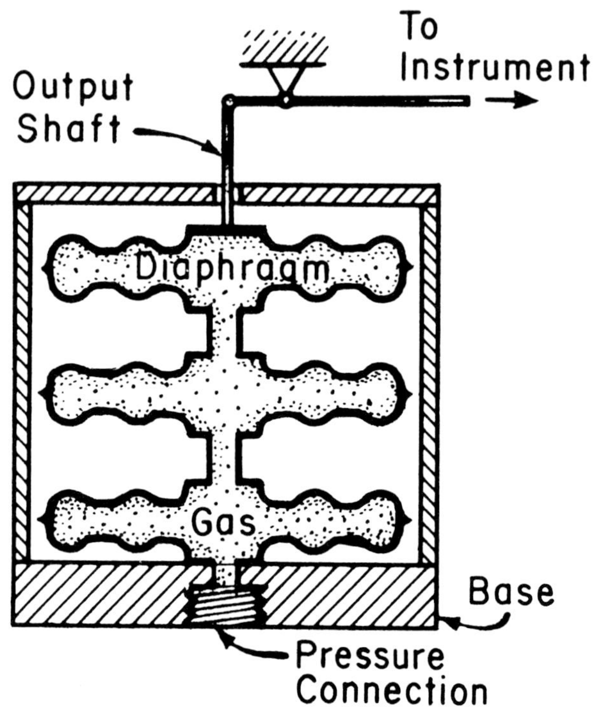

Figure 5.3. Multiple-Diaphragm Pressure-sensing Element

A *bellows* unit is similar to the diaphragm element except that the corrugations extend in series rather than expanding outward. In figure 5.4, the movement of the bellows resulting from an increase in pressure is opposed by a spring. This spring extends the range of pressure a particular size bellows is capable of measuring. Again, the movement of the bellows positions an indicating pointer, a recording pen, or the detection mechanism of an automatic controller. These units are available with ranges up to 100 lb.

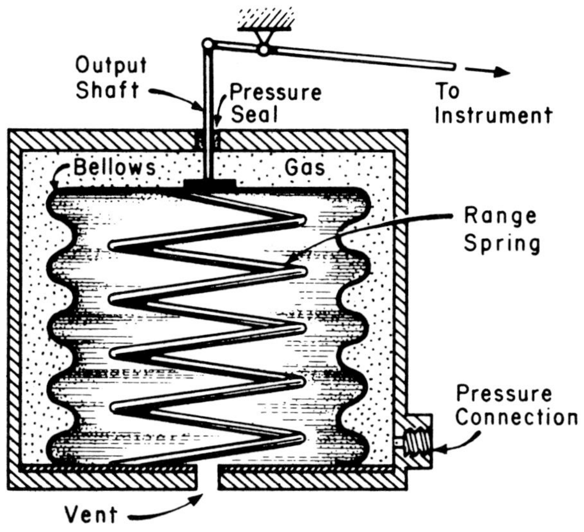

Figure 5.4. Bellows Pressure-sensing Element

A *Bourdon tube* is essentially a C-shaped tube that is sealed at one end and equipped with a pressure connection on the other end. As shown in figure 5.5, pressure introduced into the tube will apply an equal force on the entire inner surface of the tube. Since the outer circumference is greater than the inner circumference, pressure tends to straighten the tube. The movement of the sealed end of the tube is used to position a pointer, a recording pen, or an error-detecting unit. The amount of movement of the sealed end is directly proportional to the portion of a circle formed by the tube. A normal Bourdon tube has a small movement for minor changes of pressure, and this motion must be amplified with suitable linkage if a long scale is needed. This again increases calibration problems and the possibility of error. If the tube is wound in a spiral, more tip movement will result from a pressure change. If the spirals are wound in a helical shape, as shown in figure 5.6, each spiral is of the same circumference, and the accuracy of the tube is improved considerably. Each type of tube is available for measuring pressures up to thousands of pounds.

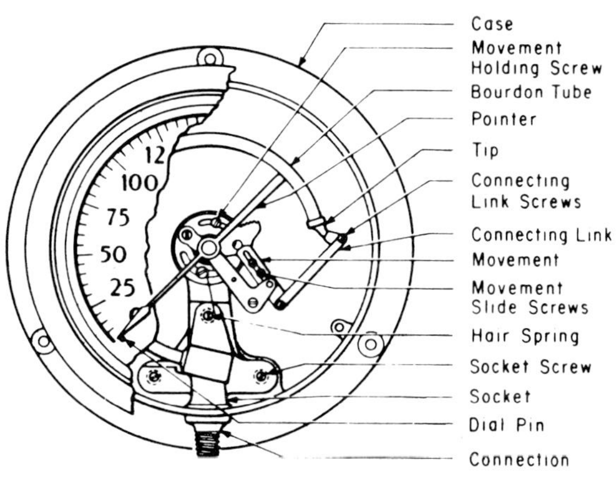

Figure 5.5. Bourdon Tube Pressure-indicating Instrument

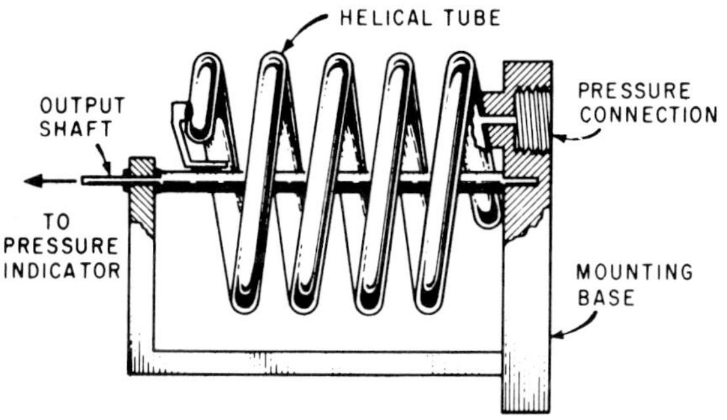

Figure 5.6. Helical Tube Pressure-sensing Element

Temperature Measurements

The common indicating *thermometer* is the most widely used instrument for measuring temperature. It consists of a bulb filled with liquid—which may be mercury, pentane, toluene, alcohol, and the like—connected to a glass stem. Changing temperature causes the liquid to expand or contract and move up or down the glass stem thus giving a visual indication of temperature. The bore of the stem is very small, and the volume of fluid in the bulb is about 1,000 times the volume of the stem bore.

In order to check the temperature of a flowing stream of gas or liquid, a method must be provided to get the thermometer bulb inside the pipe. The *thermometer well,* illustrated in figure 5.7, is permanently mounted in the piping system. When the temperature is checked, the thermometer is inserted in the well, and packing is installed in the annular space around the stem in order to eliminate variations due to ambient temperature. The well should be filled

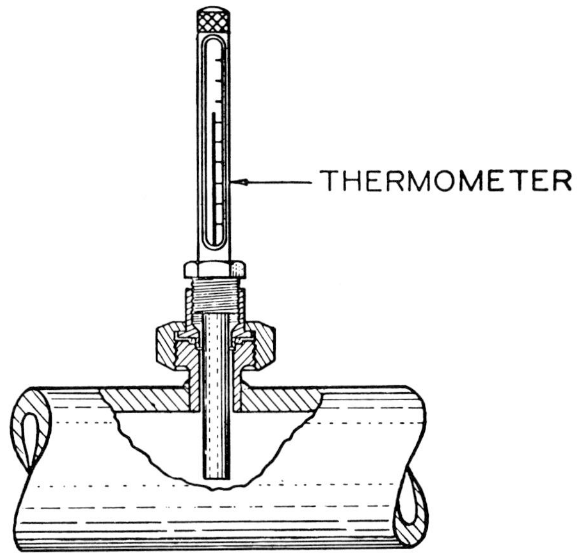

Figure 5.7. Thermometer Well

with a suitable liquid, such as mercury, in order to get good heat transfer between the material and the thermometer.

A filled-system thermometer, as shown in figure 5.8, is used to provide a mechanical movement needed to position an indicating pointer, a recording pen, or an error-detection unit in an automatic controller. The system

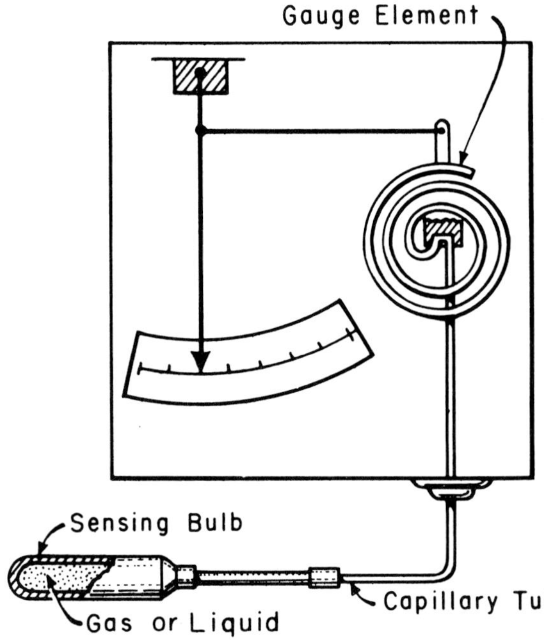

Figure 5.8. Filled-System Thermometer

consists of a bulb, small-diameter connecting tubing, and a hollow spiral tube. The entire system can be filled with a fluid such as mercury or a gas. Other systems are partially filled with a fluid that tends to vaporize at the temperature range being measured.

Systems completely filled with a fluid or gas depend directly on the expansion and contraction of the fill medium to position the tip of the spiral tube. Since ambient temperature affects the fill medium in the connecting tubing and the spiral elements, this type of system must have a compensating device. Compensation is usually accomplished by a second spiral element connected to a dead-end tube (fig. 5.9).

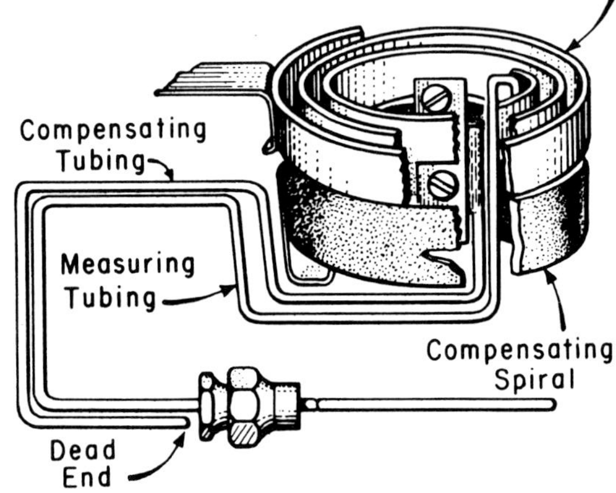

Figure 5.9. Compensating Filled-System Thermometer

The two elements are spiraled in opposite directions in such a fashion that the free end of the compensating element provides the mounting location for the measuring element. Thus, any movement in the measuring element occurring as a result of an ambient temperature change will be exactly corrected by the compensating spiral.

Systems partially filled with a fluid depend on the change in vapor pressure resulting from temperature changes to operate the spiral tube. In this type of unit, the tube constitutes a pressure-element measuring vapor pressure, and it is not affected by changes in ambient temperature. However, the measuring range for each type of fluid (ethyl alcohol, ether, propane, butane, etc.) is rather limited, and the boiling point of the fill fluid must be lower than the lowest temperature to be measured in each case.

A *bimetal temperature element* is comprised of two different metal strips. Every metal or alloy expands or contracts with temperature changes; however, the rate of expansion varies from one metal to another. If two dissimilar metal strips are bonded together along their length and held rigid at one end, the free end of the unit moves as the temperature changes. As the temperature increases, one strip of metal expands to a greater length than the other. However, since the two strips are bonded together, the entire unit must bow in such a fashion that the longer strip occupies the outer circumference of the arc. A reverse action will occur as the temperature decreases. (See fig. 5.10.) The movement of the free end of the unit is directly proportional to the length of the bimetal strip. To conserve space, most temperature-sensing strips are normally wound in a spiral or helical shape. These units are generally used as indicating devices or for safety units.

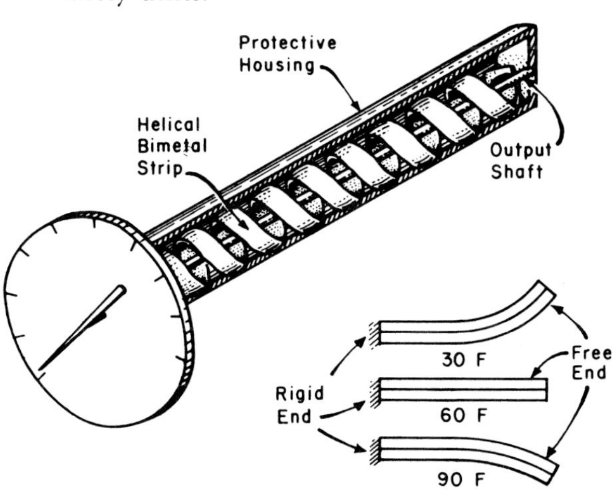

Figure 5.10. Bimetal Element

A *thermocouple element* consists of two different metals attached together. When any type of metal is heated, negative particles of electricity (electrons) are driven away from the heat. The number of electrical charges moved varies in different types of metal. If two dissimilar metals are joined at one end and the junction heated, electrical charges move down each metal and accumulate at the free ends. However, there is a greater accumulation at one free end than at the other due to differences in the two metals; the difference in accumulated charges is directly proportional to the heat applied to the junction between the two metals. (See fig. 5.11.) The

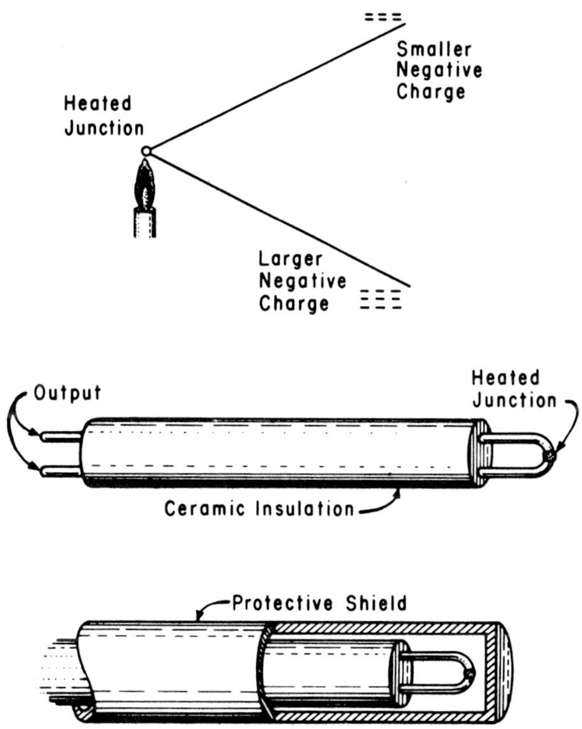

Figure 5.11. Thermocouple

difference in the electrical charge at the free end amounts to only a few thousandths of a volt, even at extremely high temperatures of 3,000 F or more. If a very sensitive voltmeter is connected to a free end, it can measure this small voltage. Even though the meter measures voltage, the scale is marked to read temperature. Thermocouples can indicate temperature directly, or the electric output can be used to energize an electric temperature controller.

Liquid-Level Measurements

The most commonly used device for a direct reading of liquid level is the sight glass. This unit consists simply of a transparent column, usually glass, suitably attached to the tank containing the liquid. When the valve between the bottom of the column and the tank is opened, fluid rises in the column to the same height as the fluid in the tank. This level can then be compared to the scale attached to the column. If the vessel is at atmospheric pressure, the top end of the column can be vented. However, on pressure vessels, both ends of the column must be connected to the tank, or fluid will be blown out the open end. Check valves must be used in conjunction with pressurized vessels to prevent fluid leakage if the glass is accidentally broken.

A *float-type unit* can be used when liquid-level control is needed. One such unit employs a bulb or ball that floats on the surface of the liquid. As the liquid level varies, the float element changes position. This motion is transmitted through suitable linkage to a shaft and through a packing gland to the outside of the vessel where it is used to position an indicator or liquid-level controller. The float assembly can be flanged directly into the vessel or into a float chamber, which in turn is connected to the vessel. The range of levels that can be measured with this device is limited since the float rides on the surface of the liquid. It is generally used for on-off dump-valve applications. Another assembly is designed in such a fashion that the liquid displaced weighs less than the bulb itself. Consequently, the float sinks in the liquid as the level rises. However, the weight of the float is reduced by an amount equal to the fluid displaced.

Figure 5.12 shows the float attached to the float-rod assembly. The float rod is held in position by a support bracket. A hollow torque tube with a sealing flange welded on one end and a suitable socket arrangement welded on the free end fits over the float-rod assembly. A small steel shaft welded to the socket assembly on the free end of the torque tube extends down the center of the tube and through the sealing flange to the outside.

When the vessel is empty, the full weight of the float causes the torque tube to twist and rotate the shaft leading outside. As the fluid fills around the float, the weight available to twist

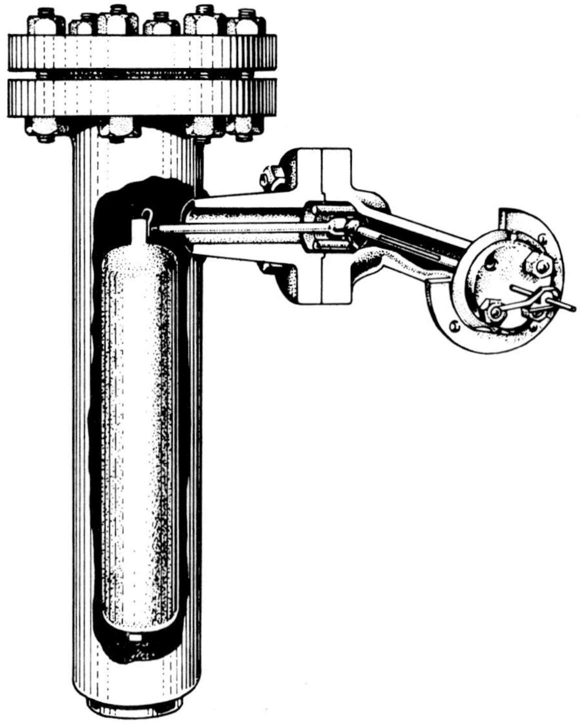

Figure 5.12. Float for Liquid-Level Controller

the torque tube is reduced by an amount equal to the fluid displaced by the float. The tube moves toward its normal position and rotates the shaft. This shaft rotation is used to supply information to the sensing device of a controller.

The force exerted by a liquid on the containing vessel is directly proportional to the depth of the fluid at the measuring point. A bellows or diaphragm mechanism can be used to measure this pressure. Figure 5.13 is a view of a diaphragm used to measure this force in a vessel vented to atmosphere. This arrangement is unsatisfactory for a pressurized vessel since the diaphragm would record not only the liquid pressure due to depth but any additional gas pressure inside the vessel as well. Figure 5.14 shows the arrangement used on pressurized vessels. The two bellows, or diaphragms, oppose each other. Both the fluid-level pressure and the gas pressure are applied to the bellows connected to the bottom of the vessel while only gas pressure is applied to the bellows connected to the top. Thus, the entire arrangement

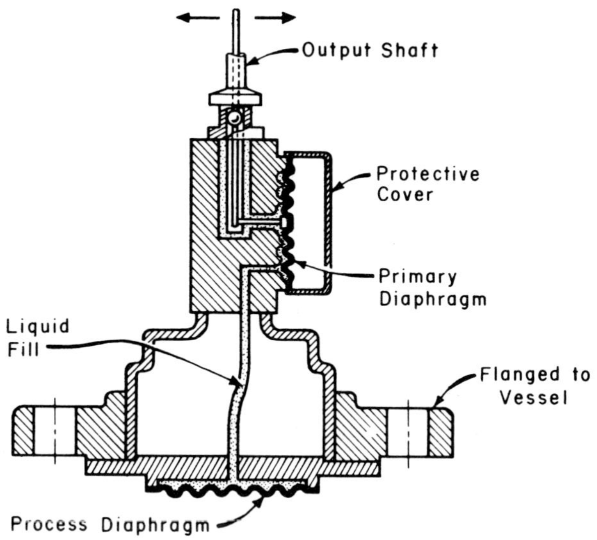

Figure 5.13. Diaphragm for Liquid-Level Controller

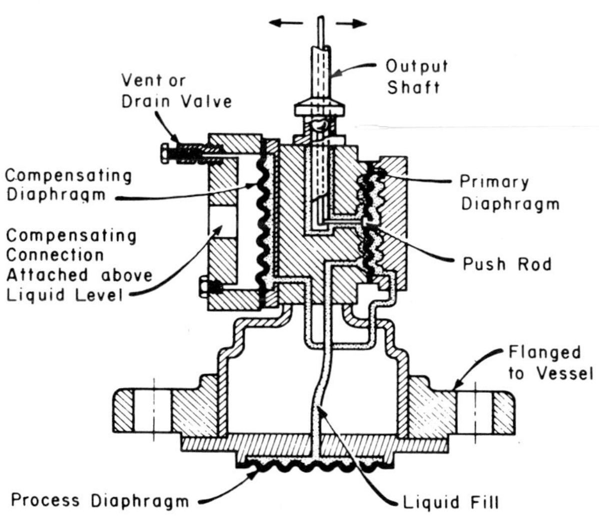

Figure 5.14. Diaphragm for Pressure-compensating Controller

measures pressure difference between the top and bottom of the vessel, which is the force produced by liquid level. The bellows movement of either arrangement is used by the error-detecting unit of a controller.

FLOW CONTROLLERS

Manual Control

Most control functions can be accomplished manually by a well-informed operator. However, all control functions can be accomplished better, more safely, more accurately, and at less cost by the proper type of automatic control. To maintain control of a process, each automatic system must contain the following essential elements:

1. a measuring element to detect any deviation in the controlled variable;
2. a controlling element that utlizes any detected error to produce a control signal; and
3. a final control valve that utilizes the control signal to position the valve at the opening required to maintain the controlled variable within acceptable limits.

The process and the automatic-control instruments are connected in a continuous closed loop giving the process the ability to control itself. Setting the acceptable limits on the controlled variable is the only manual operation required. The measuring and controlling elements are usually both incorporated in a single unit. The final control valve is usually positioned by an electric or diaphragm motor.

Flow Regulators

Every regulator, regardless of size or construction, contains three essential elements. They are (1) a restricting device to limit the flow of product through the unit, (2) a measuring device to determine the value of the controlled variable, and (3) a loading device to reposition the restricting device when necessary. A self-contained regulator contains all of these essential elements within one unit.

The *spring-loaded regulator* is most commonly used to supply fuel gas to line heaters and other burners found in the field. They are also used to supply low-pressure gas for household use. Figure 5.15 is a view of a spring-loaded regulator in its operating position. The loading element in this case is a compressed spring. This spring is applying a force to the top side of a flexible diaphragm (measuring element) causing the diaphragm to move down. This motion moves the toggle assembly and allows the seat pad to move away from the orifice.

When high pressure is admitted to the inlet of the regulator body, the gas will pass through the open orifice and into the downstream piping, causing higher pressure. This pressure increase is felt by the diaphragm since the chamber below the diaphragm is open internally to the outlet piping system. The diaphragm will move up when the force exerted by the gas pressure exceeds the spring tension. This movement closes the seat pad against the orifice to restrict or stop the flow of gas. Changing the spring tension will change the outlet-pressure setting.

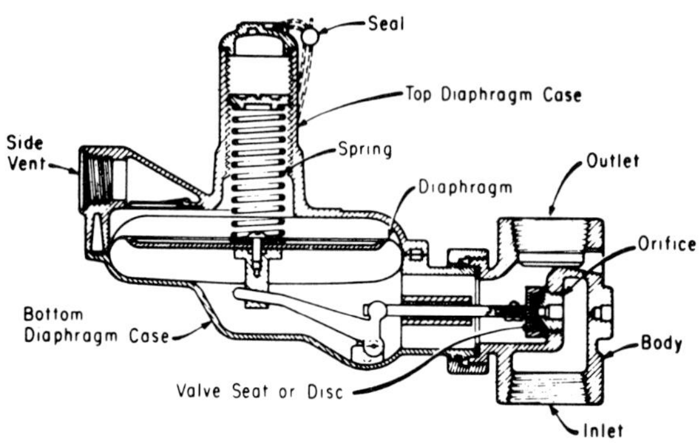

Figure 5.15. Spring-loaded Regulator

Regulators are available in this category that will accept inlet pressure above 1,000 lb and provide outlet pressure up to about 400 lb. The unit has the disadvantage of having the outlet pressure applied to the diaphragm and casing, and this is the limiting factor on maximum outlet pressure. In addition, the diaphragm must move down to open the orifice. This allows the spring to extend somewhat and relieves part of the tension. As a result, the outlet pressure drops off as the regulator opens farther.

A unit designed to control temperature is shown in figure 5.16. The diaphragm has been replaced by a bellows unit connected to a bulb. As the temperature increases, the fluid inside the bulb expands. This expansion is transmitted to the bellows repositioning the seat pad. It should be noted that this unit only controls the flow of the heating agent (gas, steam, or hot water) and totally disregards pressure. The pressure must be regulated through a separate unit.

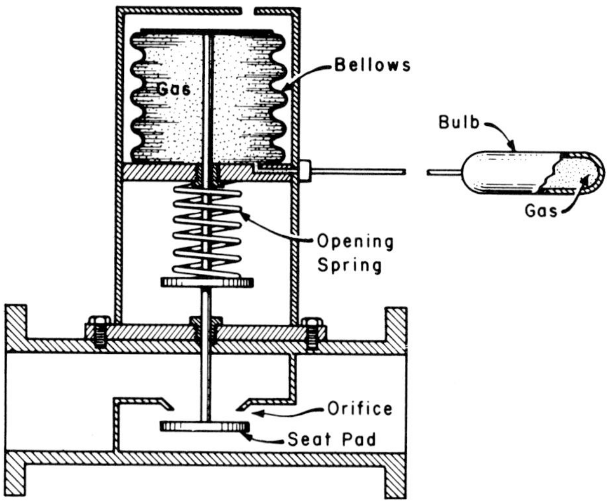

Figure 5.16. Bellows-Type Controller

Pilot-operated regulators operate on the same basic principle as the spring-loaded unit except that the loading force is a constant gas pressure rather than a spring. This constant pressure is applied to the top side of the diaphragm by a small spring-loaded pilot regulator. Most pilot-loaded units employ a closing spring to aid in holding the seat pad against the orifice. As a result, the loading pressure must overcome both the outlet pressure beneath the diaphragm and the closing-spring tension in order to open the orifice. Changing the spring tension on the pilot regulator changes the loading pressure and the outlet pressure. This unit has a constant loading force regardless of the position of the diaphragm, and this eliminates the drop in pressure associated with an increase in flow through a spring-loaded unit. However, the full outlet pressure is still applied to the diaphragm case, and this limits the maximum outlet available. These units find their greatest use in supplying gas to distribution systems and in large burner applications requiring a more stable pressure than can be delivered by a spring-loaded assembly.

Motor Valves

A diaphragm motor valve is the most common unit used for final control. Referring to figure 5.17, a valve of this type consists essen-

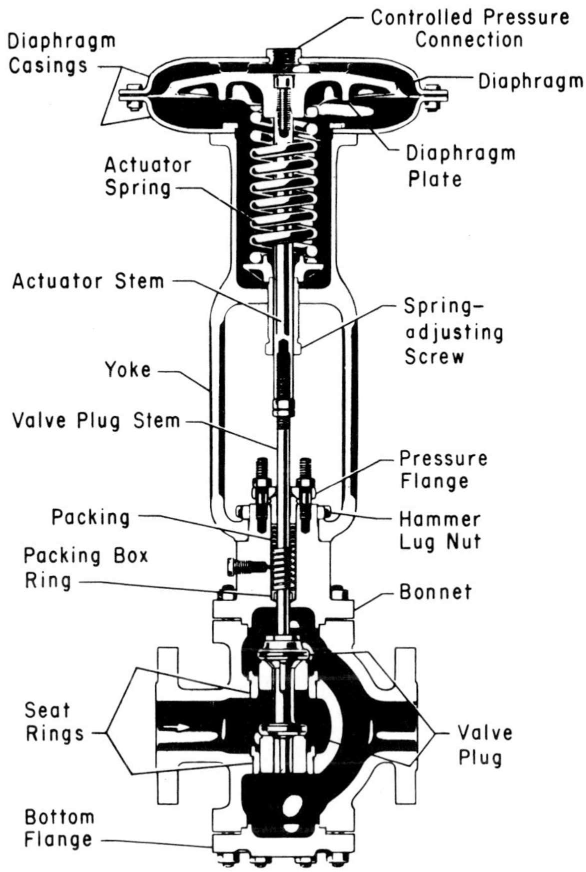

Figure 5.17. Normally Open Motor Valve

tially of a diaphragm, loading spring, valve stem, valve body, inner valve, and valve seat.

The loading spring attempts to hold the inner valve either completely open or completely closed. The first is called a normally open valve (fig. 5.17), and the latter is termed a normally closed valve (fig. 5.18). Operating pressure (control signal) is applied to the top side of the diaphragm, overcoming the resistance of the loading spring and increasing or decreasing the valve opening, depending on the arrangement of the valve seats. The pressure required to start compressing the spring is generally 3 lb; the spring then requires 15 lb of pressure to achieve maximum compression. On some valves, this pressure range will be between 6 and 30 lb. In case of complete loss of control signal, the loading spring immediately returns the valve to its normal position of open or closed.

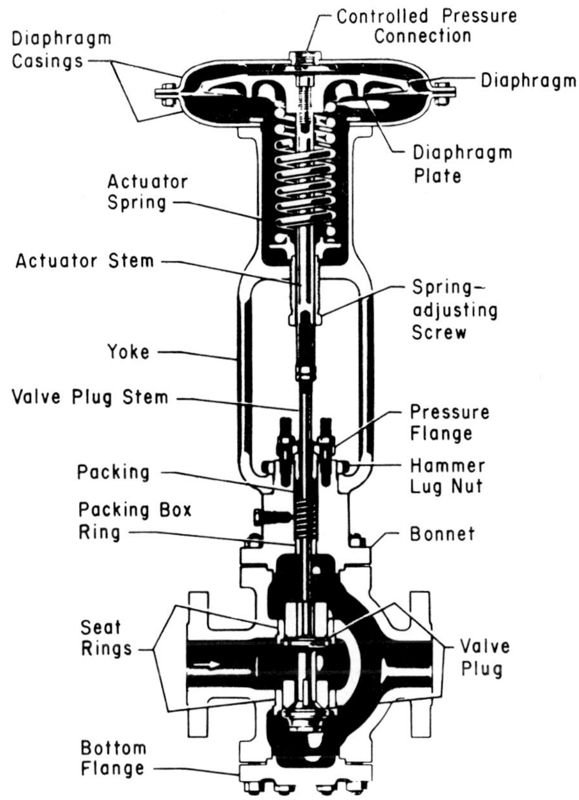

Figure 5.18. Normally Closed Motor Valve

A different arrangement for a motor valve is shown in figure 5.19. In this device, a diaphragm is still used but is attached to a hydraulic piston rather than a metal stem. An increase in the control signal moves the diaphragm down, building pressure in the hydraulic system. This hydraulic pressure is applied to a slave piston inside the body, moving the slave forward against a rubber plug. Since the plug is mounted solid at the front end, the forward movement of the slave piston deforms the rubber so that it extends to the side and contacts a smooth seating sleeve inside the regulator body. This in turn stops the flow. The flow through this body is streamlined since there is little change in direction of the flowing material. In addition, the rubber plug is capable of sealing off around trash that would normally hold other types of inner valves open. The plug will withstand a pressure drop of over 1,000 lb and still go to a complete shutoff. Unlike the conventional motor valve, the control signal range required to operate the valve depends on the pressure drop across the plug, but in most installations, it is less than the 3 to 15 lb mentioned previously.

In addition to being either normally open or closed, the inner valve of a conventional motor valve can be either single or double ported. Figure 5.20 is a single-ported valve as opposed to the double-ported valves in figures 5.17 and 5.18. The double-ported unit is sometimes referred to as a balanced valve since the material passing through the body tends to force the inner valve open at one port with the same force as it applies to the other port in a closing direction. Since the inner valve is balanced regardless of its relative position to the seat rings, the double-ported unit is used primarily when the process requires extended throttling action at less than 10 percent of capacity. However, it is very difficult to achieve a complete shutoff. The single-ported arrangement is unbalanced and not suited for service requiring a throttling of the inner valve. Since there is only one seating surface in the single-ported unit, it is used primarily when a complete shutoff is required.

Inner valves are produced in various forms, each with a particular flow characteristic in mind. Five of the most widely used valve forms are shown in figure 5.21, and the relationship between valve movement and percent of maximum flow is shown in the chart. The linear plug provides the most linear flow characteristics; that is, 20 percent of valve travel provides 20 percent of maximum capacity, 40 percent travel provides 40 percent of capacity, and so forth. This form provides good, general-purpose control. The V-port form provides excellent throttling characteristics since 40 percent of valve movement only results in very small changes in flow. This is especially beneficial when the controlled variable can be changed or upset easily. The quick-opening valve provides a large change in flow for small changes in valve position in the throttling area. This form is used when drastic changes are expected in the controlled variable or when the controlled variable is difficult to change.

Motor valves operate with equal effectiveness in either the normally opened or closed configuration. The choice between the two is determined generally by the action that can be tolerated best in the event of control-signal failure. If

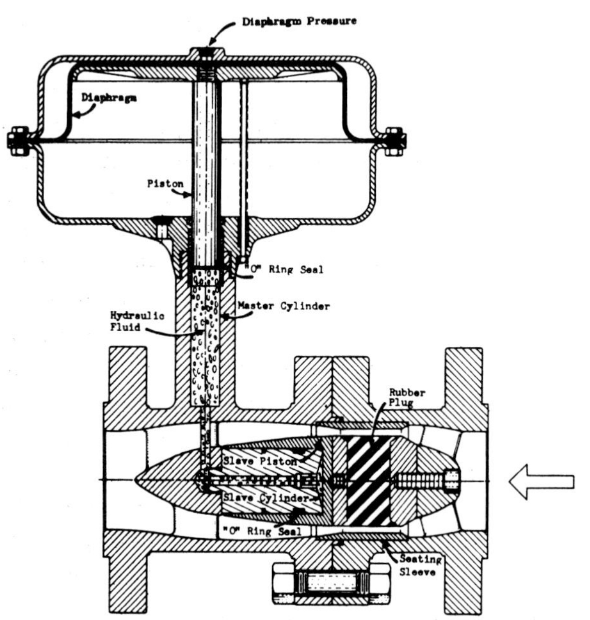

Figure 5.19. Hydraulic Piston Motor Valve

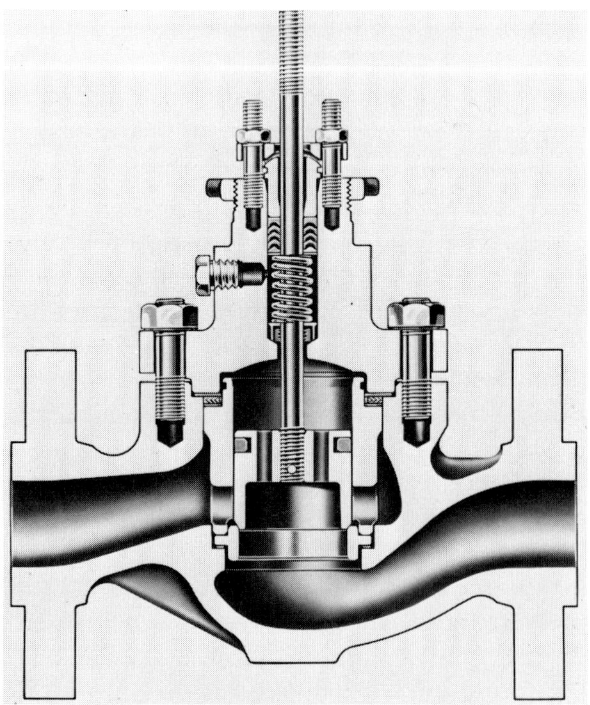

Figure 5.20. Single-Port Valve Body

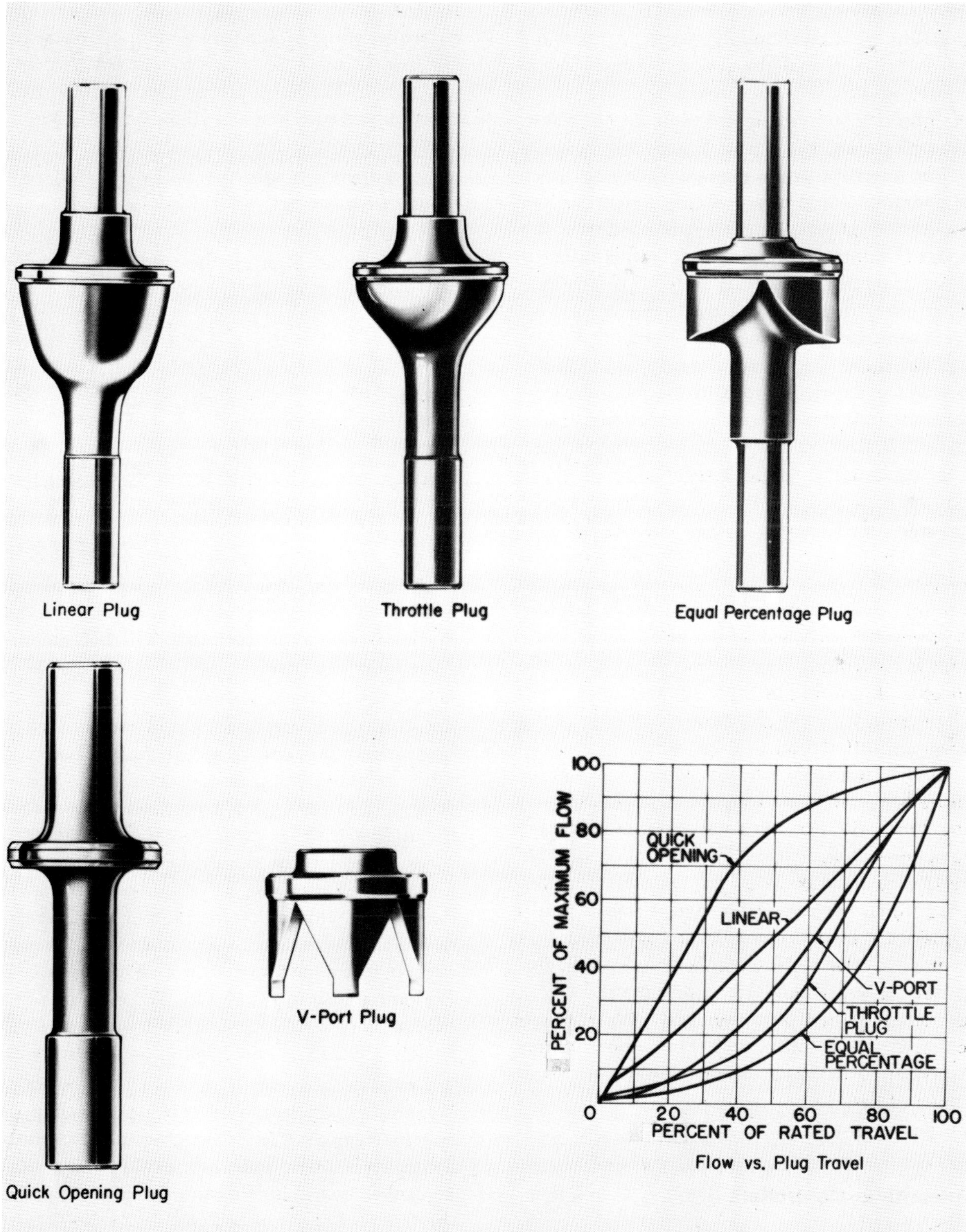

Figure 5.21. Valve Forms for Control Valves

the installation controlls pressure in a transmission line or a distribution system, a normally open valve would be used. Control-signal failure would allow the valve to go open, build the pressure somewhat, and blow a relief valve. However, customer service from the line would not be interrupted. On the other hand, this type of operation could be potentially dangerous in a processing plant. Escaping gas creates a fire hazard, and excessive pressure, temperature, or flow generally results in a loss of product. Therefore, the normally closed valve is used for most purposes in a processing plant.

The choice between a single- or double-ported valve is determined by the job to be performed and the operating conditions. Long transmission lines or distribution systems contain millions of cubic feet of storage space downstream of the valve. The valve must be capable of delivering large volumes of gas, generally is not required to throttle for extended periods of time, and never goes to a complete shutoff. This situation is ideal for a double-ported valve equipped with a linear plug. If the unit is controlling flow instead of pressure, the throttle plug would be more suitable since small changes in valve position would not change the recorded differential so drastically.

If the controlled variable is reduced two or more times in a relatively short section of piping, there will be a very small amount of storage area between the valves. The controlled variable can be changed very easily; therefore, a throttle-plug inner valve must be used if stable conditions are to be maintained. On installations that require a valve to be either wide open or completely closed (liquid level, dump valve, etc.), the single-ported unit equipped with a quick-opening plug is satisfactory. The same unit may be required for applications handling corrosive material that would erode any type of double-ported assembly.

Pneumatic Controllers

Diaphragm motor valves must be equipped with a pneumatic controller in order to automatically control any process. In addition, the controller must be furnished a supply of clean, dry, noncorrosive air or gas to be used as a control signal. The supply must be regulated to a constant pressure (usually 20 lb) before being introduced into the controller.

Controllers provide the following functions related to automatic control:

1. a method of setting the value to which the controlled variable is to be maintained;
2. a device to measure the actual value of the controlled variable;
3. a system to detect any deviation of the controlled variable from the desired setting; and
4. a system to react to the detected deviation and correct the control signal.

In most of the newer or more sensitive controllers, an amplifying device is also employed to boost the small corrective signal produced by the controller to a level that will operate the valve more effectively.

Any device that can determine the value of the controlled variable can be considered a measuring element. This includes Bourdon tubes, bellows, temperature elements, manometers, flow meters, and so forth. In order to get the most sensitive control from the equipment, a measuring element should be chosen with the lowest maximum range that will accommodate the process without having to operate at the extreme end of the range. In other words, never use a 600-lb element to control at 50 lb or a 100-lb element to control at 95 lb. The element must be constructed of a material that will not be adversely affected by the product or atmosphere.

Deviation of error detection is accomplished in most controllers by a nozzle and flapper system. Figure 5.22 is a sketch of a very simple controller using a Bourdon tube to measure the controlled variable. Suitable linkage connects the free end of the Bourdon tube to the flapper at point *A*. As the controlled pressure changes, the Bourdon tube expands or contracts and

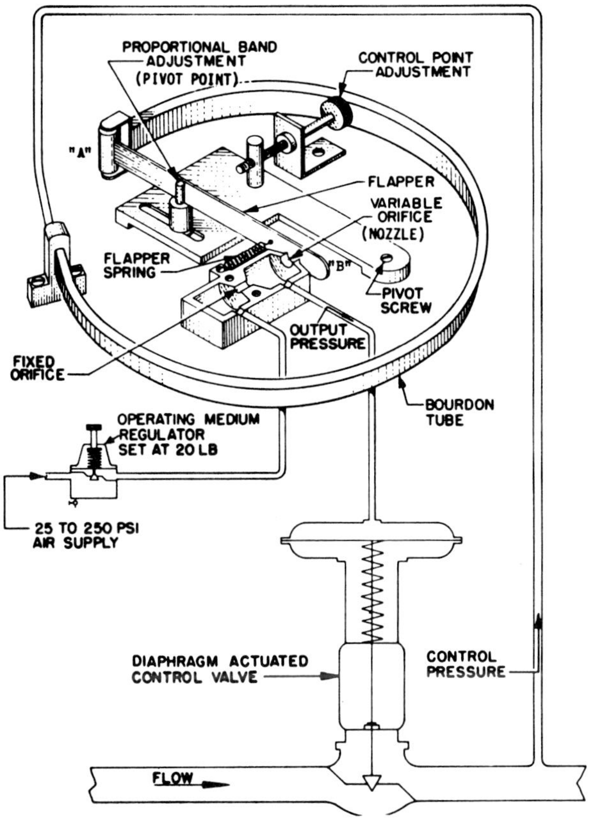

Figure 5.22. Bourdon Tube Controller

moves the flapper at point *A*. This in turn causes the flapper to rotate about the pivot point and partially open or close the small orifice at *B*. The error is determined by the distance between the flapper and nozzle. Once the error has been detected, a corrective signal is generated by two orifices, one in the supply line and one at the nozzle tip. A constant 20 lb of pressure is supplied to the primary restriction by a small pilot regulator. This primary restriction allows only a specific maximum flow into the control system. The hole in the nozzle is larger than the primary restriction; consequently, with the flapper away from the nozzle, the flow out of the control system can be greater than the incoming flow, and pressure cannot build. This control system is tied into the diaphragm chamber of the motor valve. If the valve is normally open (i.e., diaphragm pressure is required to close the valve), the inner valve will be completely open. If gas is allowed to flow through the valve, the outlet pressure will increase and cause the Bourdon tube in the controller to expand, rotate the flapper about the pivot point, partially close the nozzle, and restrict the flow out of the control system. This restriction of the flow out of the control system allows pressure to build up on the motor-valve diaphragm and close the inner valve. In normal operation, the nozzle is neither completely open nor closed; instead, the nozzle is restricted to a point of keeping the control signal at the level required to hold the valve closed enough to maintain the desired downstream pressure. The value of the controlled variable is adjusted by changing the distance between the flapper and nozzle; in most instances, the adjustment is achieved by moving either the pressure element or the nozzle assembly. If the controller is installed on a normally closed valve, the control action can be reversed; that is, an increase in the controlled variable will result in a decrease in the control signal. This reverse action will occur if the nozzle is moved to the opposite side of the flapper.

Proportional Action

When a controller is installed, what type of response is required? The answer will depend largely on the system under control; how much variation from the desired value is permissible, how much piping is downstream of the regulator, and so forth. Of course, the ideal situation would be no variation at all, but in most instances this snap-action type of control cannot be tolerated except in liquid-level control. If the controlled variable goes above or below the desired value by a slight amount, the inner valve will go completely open or closed. This results in a very erratic control pattern, and in most applications the reaction of the regulator to a change in controlled variable must be restricted. This is accomplished by adding proportional action to the controller. *Proportional action* simply means the amount of change in the controlled variable required to completely stroke the inner valve from one extreme position to the other. Proportional settings are generally given in percentage, and this denotes the percent of the total control range required to stroke the valve. In a hypothetical case, a controller could have a 100-lb spring range and a proportional setting of 10 percent. A change in the outlet pressure of 10 percent of the 100 lb (or 10 lb) would be required to stroke the valve. If the valve shuts off completely when the controlled pressure reaches 50 lb, the pressure will drop to 40 lb before the valve will completely open. Of course, at 45 lb the valve is halfway open.

In several controllers, changing the proportional setting is accomplished by moving the pivot point along the flapper. In figure 5.22, if the pivot point is moved near point *A*, a small change in outlet pressure will result in a small movement of point *A* but a large movement at the nozzle end of the flapper due to the pivot location. This in turn will completely stroke the valve. The controller is at 0 percent proportional setting. If the pivot point is moved to a position near the nozzle, a great deal of change in the controlled pressure is required to get enough movement at point *A* to bring the flapper against the nozzle. The proportional setting is now 100 percent.

Setting Proportional Action

If there could be an instantaneous repositioning of the valve whenever a deviation in the controlled variable occurred, operation at very near 0 percent proportional setting could be achieved. However, some lag will always exist since the deviation must first be detected and a corrective control signal generated before the valve is repositioned. Such things as the size of controller-restricting orifices and the volume within the control system (i.e., the connecting tubing and diaphragm case) affect the maximum speed at which the valve can be repositioned. This maximum repositioning speed will remain constant for any particular installation. The proportional setting is made in accordance with this lag time. Adjustment of proportional action is started at a high setting. The controlled variable will remain stable, but a large deviation from optimum value can exist. As the proportional setting is brought toward 0 percent, the deviation from optimum value will reduce. The adjustments are made in small increments with enough time between adjustments to allow the process to stabilize. Eventually a setting will be reached in which the required change in variable occurs faster than the valve can be repositioned, and the process will be unstable. The proportional setting is then increased enough to stabilize the system for maximum response speed.

A more elaborate control system is shown in figure 5.23. This unit is equipped with a constant supply to a primary restriction, a secondary restriction at the nozzle that is larger than the primary restriction, a measuring element, a flapper, and a pivot point. The arrangement is a little different. A bellows opposed by a spring has been added, and a piece of tubing ties the bellows into the motor-valve control system. Any increase at all in the controlled variable is enough to bring the flapper against the nozzle. As the pressure on the diaphragm builds, the change is felt by the bellows, and it extends against the spring forcing the beam down at point *B*. The beam rotates about the pivot point and lifts the flapper away from the nozzle, opposing the action of the measuring element. The amount of opposition depends on the position of the pivot point. If the pivot is near *A*,

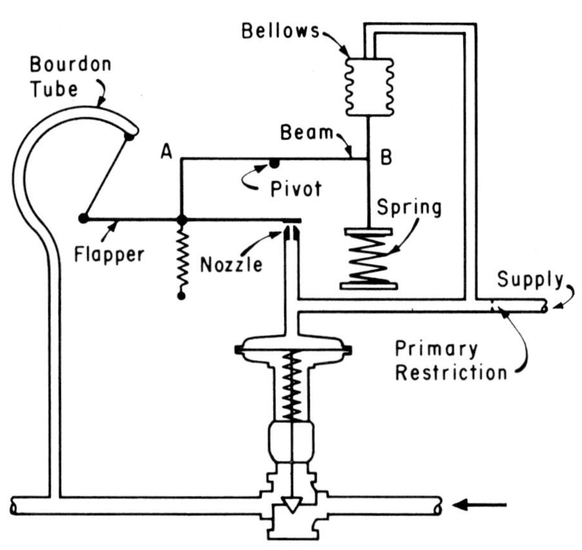

Figure 5.23. Wide-Range Proportional Controller

movement of the bellows has little effect on the flapper. Under this condition, a slight change in controlled variable is enough to stroke the valve, and the proportional setting is 0 percent. If the pivot is moved near *B*, the bellows movement causes a great deal of opposition to the flapper movement created by the measuring element, and a large change in variable is required to stroke the valve. This type of controller can be designed to go up to 400 percent proportional adjustment.

Reset Action

The controllers described previously can only have the proportional setting as close to 0 percent as possible without creating an unstable control condition. This means the controlled variable must deviate from the optimum level, depending on load conditions, in order to open the valve. In some applications, this deviation cannot be tolerated. Automatic reset allows the slightest change in controlled variable to stroke the valve completely, if necessary, and to regain the optimum value without getting into the inherent erratic control associated with normal 0 percent proportional action. In effect, reset action makes the controller operate at 0 percent proportional band regardless of the actual proportional setting. However, the controller reaction to a change in variable is retarded to prevent the erratic control.

Figure 5.24 is a control system very similar to that just described. However, the proportional bellows is opposed by a reset bellows rather than a spring. This reset bellows, along with the proportional bellows, is tied into the control system. A change in the pressure to the diaphragm is felt by the proportional bellows immediately while the change is delayed to the reset bellows by the reset valve. This reset valve is a very fine needle valve. With the pivot point moved under point *A*, the controller operates at 0 percent proportional band. Any change in the controlled variable is enough to cause the controller to stroke the valve since the feedback and

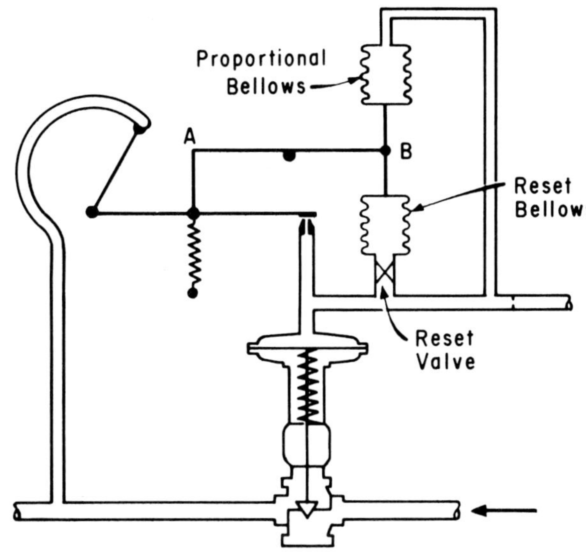

Figure 5.24. Proportional Controller with Reset

movement of the bellows cause no movement of the flapper. The same condition exists if the beam remains in one location or can be brought back to an original position. The following set of circumstances might exist with this type of controller:

1. 100-lb pressure spring;
2. 3- to 15-lb diaphragm-pressure change required to stroke motor valve;
3. 10 percent proportional setting;
4. 1 repeat per minute setting on reset valve; and
5. system stable with 14 lb on entire diaphragm-pressure system including the proportional and reset bellows, making the beam completely level at this time.

If due to additional consumption the controlled variable drops off 1 lb from the ideal value or 1/10 of the amount required to stroke the valve using proportional action alone, the pressure element will cause the flapper to move away from the nozzle and vent diaphragm pressure. This lower diaphragm pressure is felt immediately by the proportional bellows but not by the reset bellows because the reset valve is almost closed. Consequently, the pressure in the reset bellows remains at the original pressure temporarily, acts as a spring, forces the beam up at point *B* and down at point *A*, and opposes the action of the pressure element. Under the circumstances stated in the example, there would be a net drop in diaphragm pressure initially of 1.2 lb due to proportional action alone. So far, the response is the same as encountered in a controller with proportional action only. In that case, the proportional bellows was opposed by a spring that applied a constant force. In this application, the entire diaphragm-pressure system, with the exception of the reset bellows, has dropped from 14 lb to 12.8 lb. There is a 1.2 lb difference in pressure between the two bellows due to initial proportional action. Without reset action, the controller would maintain this 12.8 lb on the control system as long as consumption remained at this new rate.

The pressure gradually bleeds out of the reset bellows. As this pressure bleeds out, the force exerted against the proportional bellows decreases, and the beam moves down at *B*, moves up at *A*, and pulls the flapper away from the nozzle. This, in turn, reduces the pressure in the rest of the diaphragm system, and the same 1.2-lb unbalanced pressure condition exists between the reset and proportional bellows. Proportional action can either increase or decrease this unbalance if the control variable changes again. The pressure in the diaphragm system will continue to drop at a rate determined by the reset valve. In this case, the reset valve is set for 1 repeat per minute; that is, the diaphragm pressure will continue to drop a specific amount each minute, the exact amount being equal to the unbalance between the two bellows. Remember, the amount of this unbalanced condition is determined by the proportional setting. The reset action, which is an unbalanced condition between the two bellows, continues to drop the diaphragm pressure and open the valve. The pressure in the reset bellows is attempting to equalize with the pressure in the rest of the diaphragm-pressure system but is unable to do so because of the restriction at the reset valve. As the diaphragm pressure falls off, the inner valve opens more, and eventually the control variable starts to increase. As it builds, proportional action causes a change in diaphragm pressure that tends to correct the unbalanced condition created by the original drop in the control variable. When the controlled variable builds to the original value, proportional action will have exactly corrected the unbalanced condition between the two bellows. The pressure in both bellows will again be equal but not necessarily at the original 14 lb. Consequently, the beam will be back in its original position. In this type of controller, the proportional action only creates an unbalanced condition in the controller when the control variable deviates from the optimum value and provides a reverse balancing correction as the optimum value is restored. The lower the proportional setting or the larger the deviation, the greater the unbalancing signal. Once an unbalanced condition exists, the reset action continues to change the diaphragm pressure, until it strokes the valve completely if necessary, in order to restore the original value. This occurs even for a small deviation.

Rate of Response

Many processes are of such a nature that any deviation of the controlled variable outside certain narrow limits results in unacceptable products. Any controller equipped with reset action returns the controlled variable to the optimum value, but it can do so only at a specific maximum speed. If there is a sudden drastic change in the controlled variable, reset action requires a considerable period of time to restore the optimum value. In the meantime, the product is lost or ruined. Rate of response partially eliminates this problem by allowing the controller to operate at 0 percent proportional action temporarily. This is accomplished by adding a needle valve in the line supplying the proportional bellows. With this valve partially closed, feedback to the flapper from the bellows system is delayed, allowing the controller to operate at 0 percent proportional action. As the pressure in the proportional bellows gradually changes through the partially closed rate valve, normal proportional and reset action occurs. The rate valve must never be closed more than the reset valve; otherwise the controller operates at 0 percent proportional action constantly.

Controllers that rely on the restricting orifice, nozzle, and flapper alone to produce the 3- to 15-lb control signal needed to operate a motor valve generally prove unsatisfactory for anything except the on-off system or those having a very slow change in the controlled variable. Air relays almost invariably form an integral part of the sensitive controller. The back pressure that builds when the flapper moves closer to the nozzle is applied to the air relay that in turn supplies the 3- to 15-lb signal for the motor valve. In many instances, the back pressure at the nozzle needs only to vary 1 lb in order to produce the change from 3 to 15 lb being supplied by the air relay to the motor valve. A no-bleed relay is shown in figure 5.25. The only time the exhaust valve opens is when the diaphragm pressure must be reduced. Other versions of the relay have a throttling valve and vent to atmosphere constantly. The top diaphragm is approximately 12 times as large in area as the lower diaphragm. The two are connected by a metal bracket so that any movement of one causes an equal movement of the other. An atmospheric vent is connected between the two. The constant 20-lb supply is introduced to the primary restriction and to the valve in the relay. If the nozzle back pressure is increased, the diaphragms are forced down against the relay valve. Since the valve is rather large, pressure builds on the motor-valve diaphragm

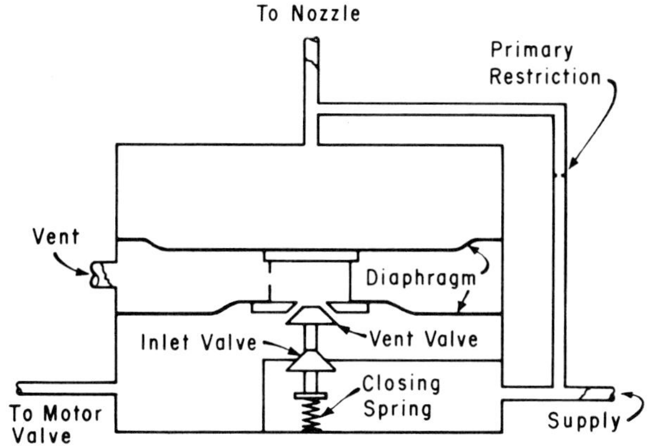

Figure 5.25. No-Bleed Relay

quickly. Due to the ratio between the areas of the two diaphragms, approximately 12 times as much pressure builds under the bottom diaphragm before the force exerted equals the opposing force created by the nozzle back pressure on the top diaphragm. When the two forces are equal, the spring under the relay valve forces the valve closed and trap pressure on top of the motor-valve diaphragm. If the controlled variable later changes so that nozzle back pressure is decreased, the greater pressure under the bottom relay diaphragm forces the assembly up and opens the vent. When enough of the pressure has been vented to atmosphere, the forces acting on the relay diaphragms are again equal and the vent closes. The advantage in using a relay is obvious. The motor-valve diaphragm is pressurized and depressurized through large openings rather than through the small controller-restricting orifices. In addition, the nozzle back pressure only changes 1 lb instead of 12 lb to stroke the valve.

The stem in a motor valve passes through a packing gland that prevents gas leakage around the stem. In some installations, this packing offers substantial resistance to stem movement, and the inner valve may not respond to small changes in the control signal. As the name implies, a valve positioner adjusts the valve to the proper opening as determined by the control signal regardless of what pressure it must apply to the motor-valve diaphragm to do so. If, for instance, the control signal is 9 lb (midway between 3 and 15 lb), the inner valve should be halfway open. The valve positioner will not be satisfied until the inner valve is in the correct position even though it may have to apply something other than 9 lb to the motor-valve diaphragm in order to get the proper response. Figure 5.26 is a sketch of the basic operating parts of a valve positioner. In an example of a normally open valve, a ¼-lb increase in the control signal applied directly to the motor-valve diaphragm might not be enough to overcome the resistance offered by the packing. The inner valve would not move. However, with a valve positioner, the control signal is applied to a

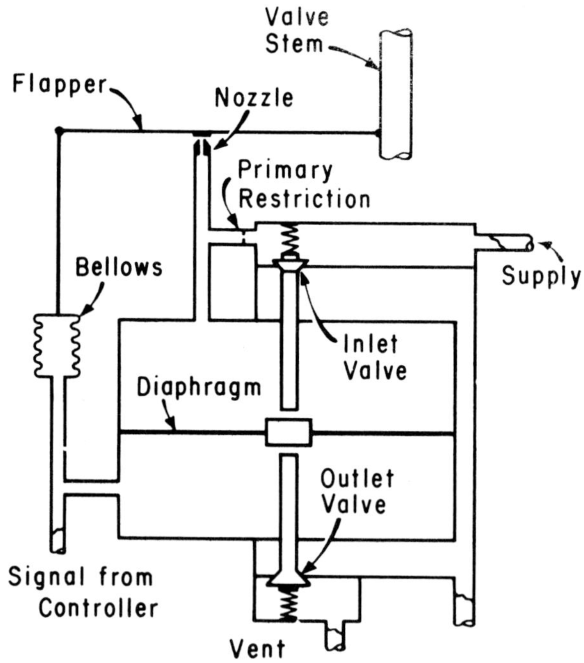

Figure 5.26. Parts of Valve Positioner

bellows and to the bottom side of a relay diaphragm. The slight increase in control signal causes the bellows to expand and lift one end of the flapper away from a nozzle. The other end of the flapper is attached to the motor-valve stem. As the flapper moves away, the nozzle back pressure decreases. This back pressure is being applied to the top side of the relay diaphragm, and any decrease allows the diaphragm to move against the supply valve. Opening the supply valve allows pressure to increase on the motor-valve diaphragm. The pressure continues to increase until the stem moves down and brings the flapper back against the nozzle. As the nozzle is closed, the back pressure to the relay diaphragm increases until it equals the control-signal pressure under the diaphragm. The diaphragm moves to the center position and allows the relay supply valve to close, trapping the required pressure on the motor-valve diaphragm. A reverse action occurs when the control signal decreases a slight amount. The relay diaphragm moves down, opens the relay vent valve, and vents pressure from the motor-valve diaphragm until the stem is moved to the position called for by the control signal. The valve positioner is usually employed on large motor valves due to the amount of static friction associated with massive packing glands. A high-pressure drop across a valve occasionally vibrates the inner valve excessively. A valve positioner aids in stabilizing this vibration.

Electric Controllers

An electric controller valve makes use of a reversible motor. Given a reversible motor, there are numerous methods of converting the rotation of the motor into linear movement suitable for operating a valve. The problem remaining is the control of the motor. Figure 5.27 is a diagram of a controller that operates the motor if the controlled variable is outside fairly wide limits. A mercury switch acts as a single-pole, double-throw switch capable of assuming an off-center position. Almost any type of measuring element is capable of tilting the mercury switch. As an example, suppose the system is used on a valve supplying fluid to a tank where fluid level is the controlled variable. If the fluid level increases to the limit adjusted into the controller, the measuring element tilts the switch enough to make contact between ter-

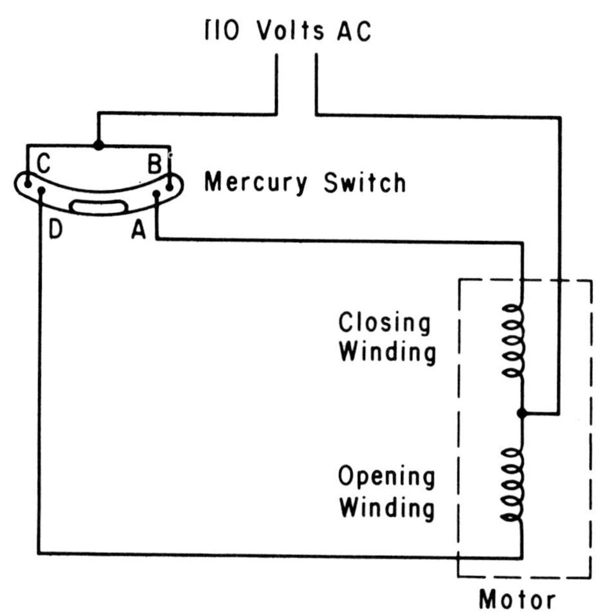

Figure 5.27. Electric Valve Controller

minals A and B, causing the motor to turn the valve in a closing direction. The motor continues closing the valve until the fluid level drops enough to open terminals A and B. If the fluid level drops too much, the opposite action occurs when contacts C and D close. Obviously, at any fluid level between the two limits, the motor is not energized and the valve opening remains at a set position. There are many applications for which this controller is not suitable due to the wide allowable deviation of the controlled variable between the two limits.

A circuit diagram of a proportional control system used in conjunction with a reversing motor is shown in figure 5.28. The valve opening in this system is adjusted for each change in the controlled variable, and the valve opening is proportional to the deviation from optimum value. The secondary from a transformer is applied to the movable contacts at terminals A and F. The arm at A is positioned by the measuring element in the controller and moves each time the controlled variable changes. The arm at F is positioned by the reversing motor and moves each time the motor is energized. These two movable arms are part of two identical variable resistor systems, and their position determines the resistance in each branch of the parallel circuits. The first circuit is from A through C into coil C_1 then through D to F. The second circuit is from A through B into coil C_2 then through E and F. The magnetic force about each coil is identical, and there is no movement of the center contact between the coils. If a decrease in the controlled variable drives movable contact A toward B, the total circuit resistance through coil C_2 decreases, the current through C_2 increases, the magnetic force around C_2 increases, and the center contact between the coils moves to terminal R. This energizes the motor circuit that opens the valve. As the valve opens, movable contact F is driven toward D. This adds resistance to the circuit through C_2 and removes resistance in the circuit through C_1. Eventually, the resistance through the parallel coil circuits becomes equal, the current flow through each becomes equal, the magnetic field about each becomes equal, and the center contact between the coils moves to the off position. A reverse action occurs when the controlled variable deviates in the opposite direction.

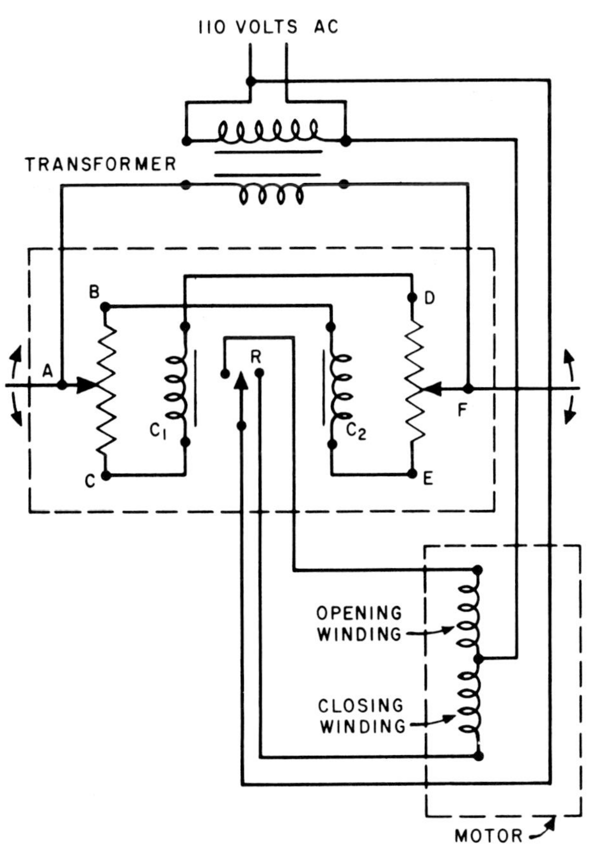

Figure 5.28. Diagram of Proportional Control System

CONTROL METHODS

At most plants, operating efficiency and product purity are maintained by the control of flow rates, liquid levels, pressures, and temperatures. In essentially all situations, the control of these variables is accomplished through the use of regulating valves.

The control of valves can either be done by an operator or by an instrument called a controller. In either case, a desired value (set point) for the variable being controlled must be established. The variable must also be measured so that when a process upset occurs it is possible to sense that the variable has changed from the desired value. The valve is then opened or closed to change the process variable. Measurement of the variable is continued so the effect of the change can be observed or sensed by the controller. The following steps are necessary to control a process variable.

1. The process variable is *measured.*
2. The measurement is *compared* to the set point.
3. If there is a difference, the *regulating valve* is adjusted.
4. *Feedback* tells the controller what has happened.

Flow rates are generally controlled by changing the output of a pump or compressor. Figure 5.29 shows three methods to control pump output. The first illustration shows an electric-motor-driven centrifugal pump. In this case, the control valve regulates flow by changing the pump discharge pressure. Figure 5.30 shows a typical performance curve for a centrifugal pump. This curve has been simplified by showing only discharge pressure on the left side. In this example, it is assumed that there is no change in the suction pressure or the specific gravity of the liquid being pumped. If the pump is operating along dashed line *A*, and the set point on the controller is changed to reduce the

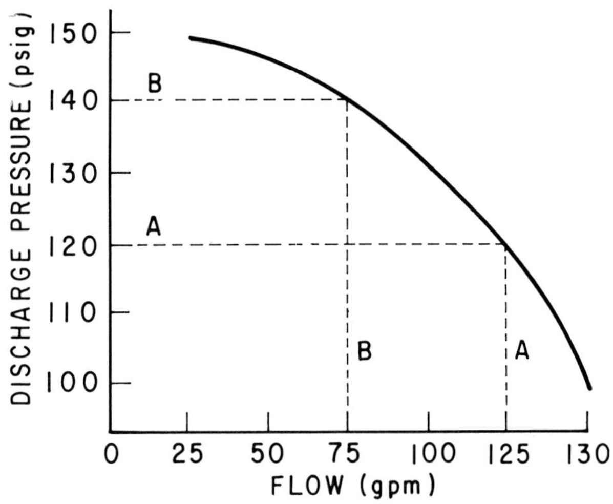

Figure 5.30. Centrifugal Pump Performance Curve

flow, then the control valve will partially close. This increases the pump discharge pressure so that the desired flow as shown as dashed line *B* is attained. In the second pump system, the flow output of the reciprocating pump is regulated by bypassing flow from the discharge to the suction. Even though the pump output is constant, the flow to the process can be varied by changing the amount bypassed. The flow from a steam-driven pump can be regulated by adjusting the steam pressure available to the driving piston arrangement.

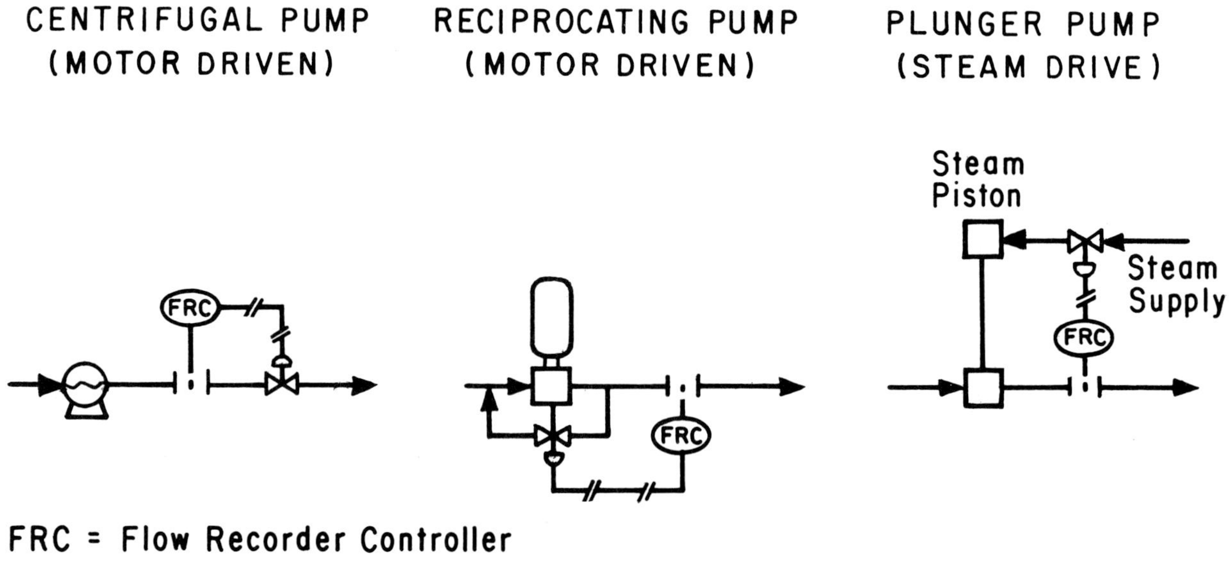

Figure 5.29. Methods of Controlling Flow Rates

Levels are controlled by the use of level controllers operating a control valve. Figure 5.31 shows that levels can be controlled by regulating the amount of liquid entering a vessel or by regulating the amount of liquid leaving the vessel. The sensing device may be a float or diaphragm arrangement (LC) to detect and indicate the fluid level in the vessel.

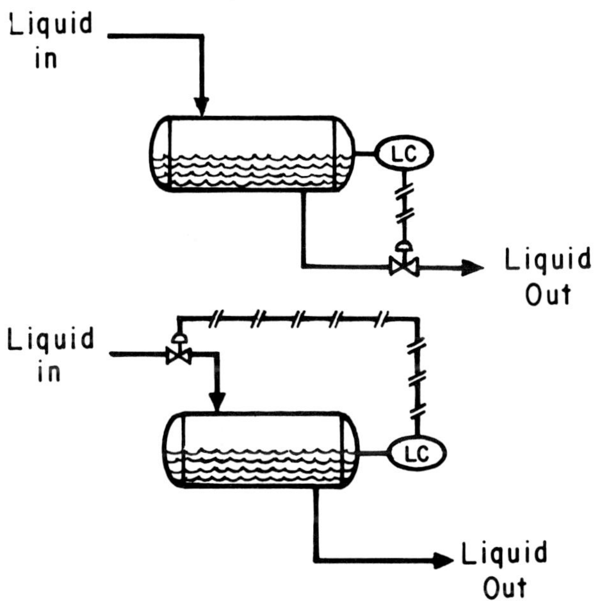

Figure 5.31. Methods of Controlling Liquid Levels

Pressures are generally controlled by regulating the amount of vapor leaving a vessel. Figure 5.32 shows the pressure control of a rich-oil flash tank by utilizing a pressure recorder controller (PRC) to open or close a valve that controls the pressure in the vessel.

In most cases, process temperatures are controlled by regulating the amount of heat added to the process stream. Figure 5.33 illustrates a method to control the heat input to a shell-and-tube exchanger by means of a temperature recorder controller (TRC) to regulate the flow of hot oil or steam through the exchanger.

Also shown is a fired heater that is regulated to control the heat input to the heater by a temperature recorder controller (TRC) that controls the amount of fuel gas going to the heater.

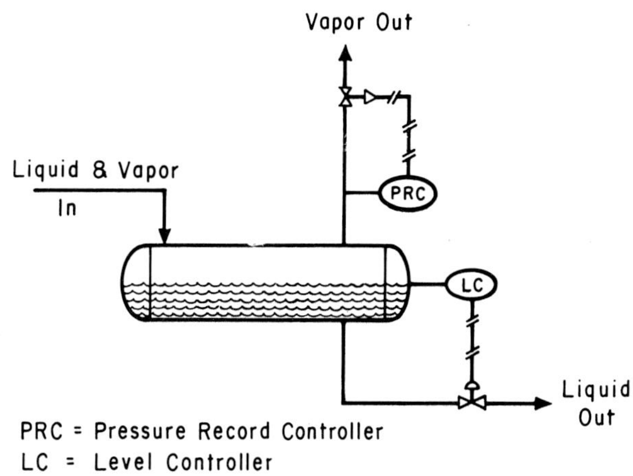

Figure 5.32. Controlling Process Vessel Pressure

Process control efficiency is influenced by system response. System response is the time required to adjust process variables to the set point. This time lag is caused by:

1. the time required for the parts of the control loop to detect a process change, make a valve adjustment, and get process feedback and
2. the time required for the process to adjust to the instrument changes.

Good instrumentation requires that the controllers respond to a change caused by either the process or the operator with a minimum disturbance to the rest of the plant. There are three devices that may be included within the controller unit to assist in achieving good instrument response and control. These devices are proportional, reset, and rate controllers, which can also be called derivative or preact controllers.

A proportional controller regulates the position of the control valve in accordance with how much the measured variable changes from the set point. A proportional controller has a range of the process variable that it controls. This range is called proportional band and is expressed as a percentage. Level controllers normally have only proportional control. Refer to figure 5.34. With 100 percent proportional band, the liquid level would have to rise to line *A* before the controller would completely open the control valve. Also, the liquid level would have

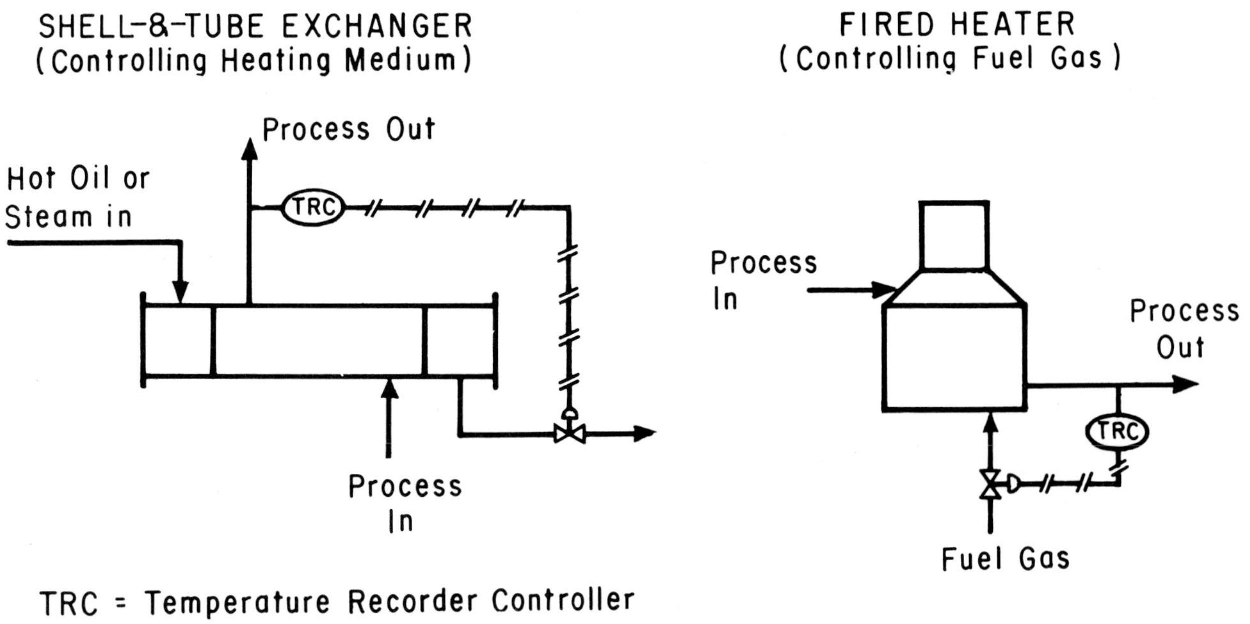

Figure 5.33. Controlling Process Temperatures

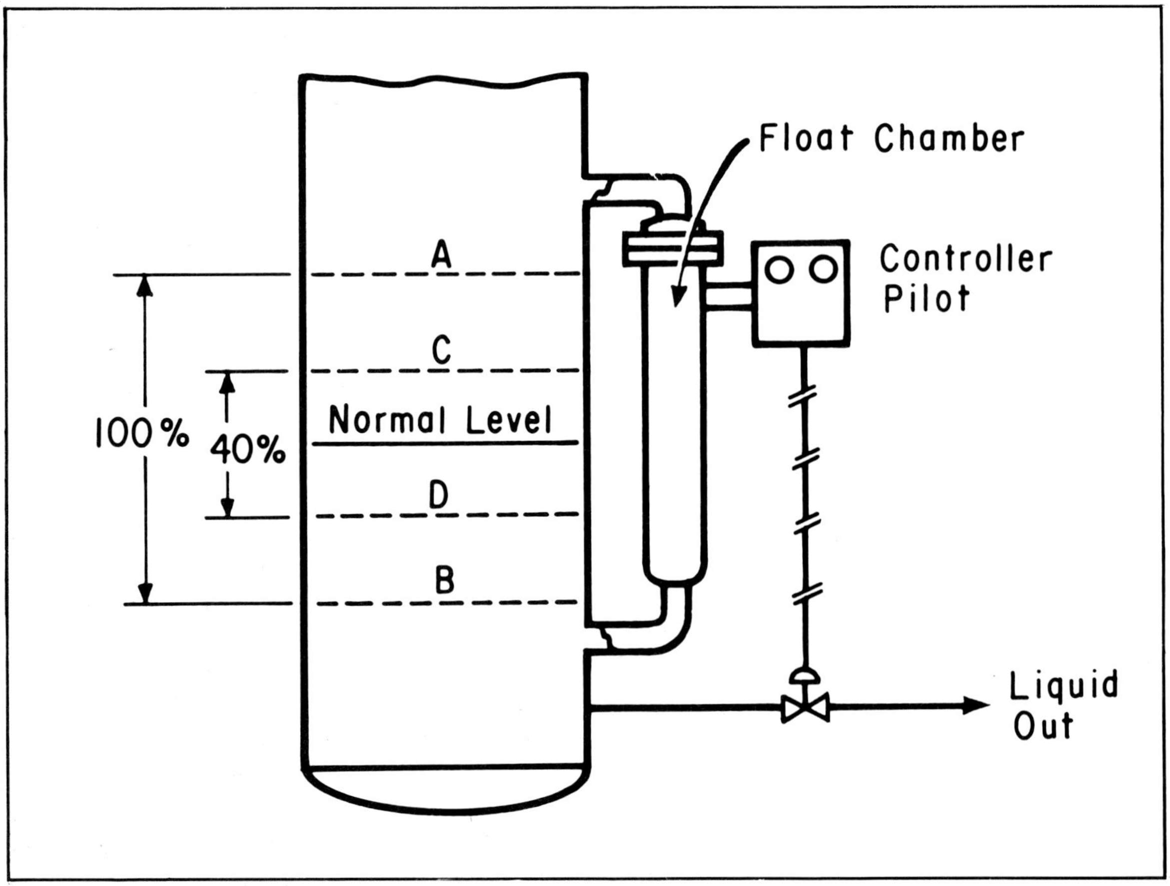

Figure 5.34. Proportional Control by Liquid Level Controller (LC)

to drop to line *B* before the controller would have the control valve closed. At 40 percent proportional band, lines *C* and *D* represent the open and closed positions of the control valve. The most stable operation is generally attained with a large proportional band, but liquid level changes from the normal will also be greater. If wide liquid level changes are unsatisfactory, then the proportional band must be reduced.

One problem with proportional controllers is that when changes occur, the controlled variable may not be returned to the set point. This difference between the measurement of the controlled variable and the set point is called *offset.* The wider the proportional band, the greater the offset when there is a load change. Offset can be overcome by adding a reset function to the controller. This is usually done for pressure and flow controllers. Reset control action causes the controller output to continue to change as long as the process measurement value is not at the true set point of the controller. The reset function does not bring the process variable back to the control point immediately but only over a period of time. The reset function is usually adjustable to allow the selection of optimum reset for each application. The more reset used (i.e., smaller time setting), the more rapidly the process variable is returned to the set point and the more rapidly the offset is eliminated. However, the use of reset tends to make a system less stable, and when used in excess can cause cycling of the system.

System response may be reasonably fast for flow rate, level, and pressure controllers, but the time lag for temperature adjustments is usually quite long. Because of this long time lag between adjustment, response, and return to the set point, there can be wide swings in temperature. To minimize this, many temperature controllers have rate action also. The part of a controller's output signal generated by rate action is based entirely upon how fast the controlled variable measurement departs from the set point. The rate adjustment acts as a temporary reduction of the proportional band. This reduction produces a very large controller output for any rapid change in the measurement. The effect gradually decreases after the measured variable becomes constant.

Cascade control is a system in which two measured variables are interlocked in one control loop. The objective of cascade control is to provide precision results by overcoming lags in a given system. This is accomplished by using the output of a primary controller to adjust the set point of a secondary controller on a related variable. Figure 5.35 illustrates the use of cascade control on the bottom of a fractionator.

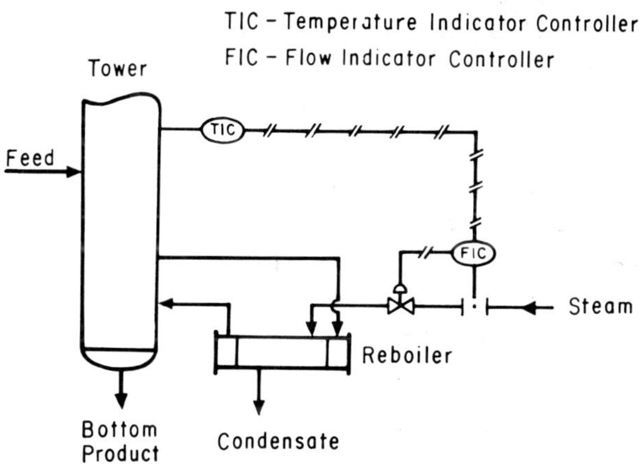

Figure 5.35. Arrangement for Cascade Control

Changes in operating conditions require an adjustment in the heat input. In order to sense a change rapidly, a temperature controller is located at an optimum control point on the tower. The output from this temperature controller adjusts the set-point of the steam flow controller. Therefore, the steam flow rate changes with a changing tower temperature. The temperature controller is the primary controller. The steam flow controller is the secondary controller.

Process control systems can generally be designed so that the system response is adequate. However, there are times when the operator must help the control loop by taking action, such as opening bypasses. Unless an unsafe condition is occurring, set-point adjustments should be made in small increments. The operator should give the process time to "settle" before making the next adjustment.

CONTROL SYSTEMS

To a large degree, instrumentation is a matter of personal preference. However, the selection of a process control system should be based on the following considerations:

1. make sure the process is controllable;
2. make sure the measurement is feasible, representative, and made at the proper location;
3. for simplicity and economy, minimize the equipment required;
4. consider emergency and unusual operating conditions and provide for "off-design" operation; and
5. consider any material and heat balances that must be made by the controls.

Process instrumentation can best be described by using examples of some actual installations. Most of the principles involved in process control systems appear in the instrumentation of either an absorber or a fractionator.

Absorber

Figure 5.36 shows one method of control for a refrigerated absorber. A flow indicating controller (FIC) regulates the lean-oil flow to the absorber. The level controller (LC) at the bottom of the absorber regulates the rich oil leaving the absorber. A low-level alarm (LLA) is installed near the bottom of the absorber. If the level controller or its control valve should malfunction, inlet gas could pass through the control valve and overpressure the downstream facilities. This low-level alarm would be either a float type or a differential pressure sensing device. In either case, the actual rich-oil liquid level would be sensed.

The residue gas scrubber has a level controller. For this refrigerated absorber system, liquid would not normally be present in the residue gas scrubber. If an absorber upset occurred, lean oil would probably be carried overhead with the absorber residue gas. In this event, when the level controller output reached

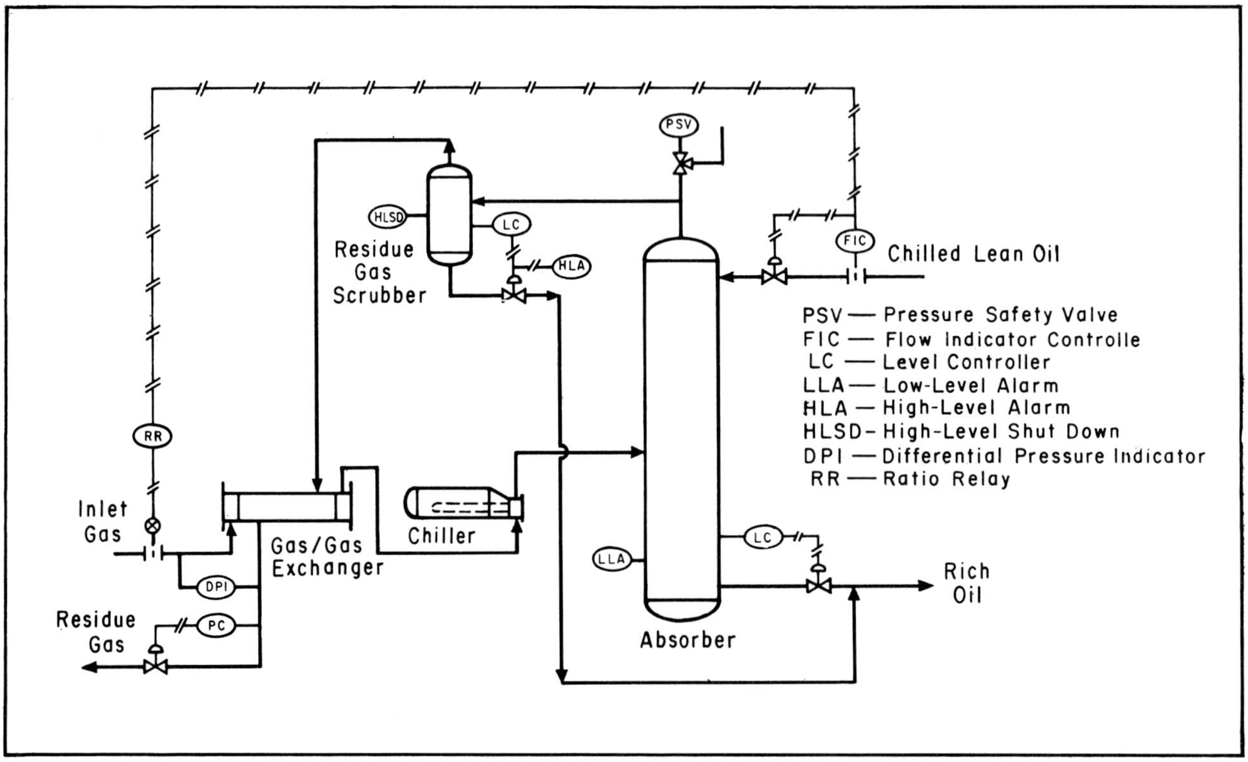

Figure 5.36. Instruments and Controls for a Refrigerated Absorber Unit

about 6 psi (indicating that the level control valve was opening), a pressure-switch-type, high-level alarm would sound. This high-level alarm does not indicate the actual liquid level in the residue gas scrubber; rather, it only signals that the level controller is in operation. Knowing that the level controller is in operation is good because this is an indication that an upset is occurring in the absorber. A float-type, high-level alarm could be used, but it is more expensive than the pressure-switch type. Also, if the carryover is small, the level controller may be able to handle the liquid, and there would not be any buildup of liquid level.

A high-level shutdown (HLSD) is also installed on the residue gas scrubber. This should be a float-type device since the actual liquid level is the main concern. If the liquid level becomes too high, then either the plant is shut down or the lean-oil flow to the absorber is stopped. Otherwise, lean oil would flow out of the plant with the residue gas.

Absorber pressure is maintained by a pressure control valve installed on the residue gas line. If the pressure exceeds the material design pressure, a pressure safety valve (PSV) will open and relieve the excess pressure.

Some plants, particularly those that are unattended during part of the day, have a control system for regulating the lean-oil flow in proportion to the inlet gas flow. The ratio relay (RR) changes the set point of the lean-oil flow indicating controller. This control is particularly useful when changes in inlet gas volume are frequent and substantial.

The differential pressure indicator (DPI) allows an operator to detect freezing problems within the system. An increase in the pressure drop between the inlet gas and the residue gas indicates either an increase in the gas volume or blockage of the gas flow path.

Fractionator

Temperature and flow control of fractionators are covered in chapter three. This discussion will be limited to some of the various methods of pressure control. Figure 5.37 illustrates a *hot-gas* bypass. Pressure controller *A* maintains the tower pressure. However, the accumulator pressure varies according to the condenser outlet temperature. This can be seen in figure 5.38. A decrease in condenser outlet temperature will lower the accumulator pressure because the lower temperature reduces the vapor pressure of the overhead product. Without any control, the accumulator pressure could become so low that the reflux pump would not be capable of pumping reflux back to the tower. To avoid this problem, pressure controller *B* (fig. 5.37) is used to bypass hot gas from the tower overhead around the condenser. This controls the accumulator pressure by maintaining the accumulator temperature.

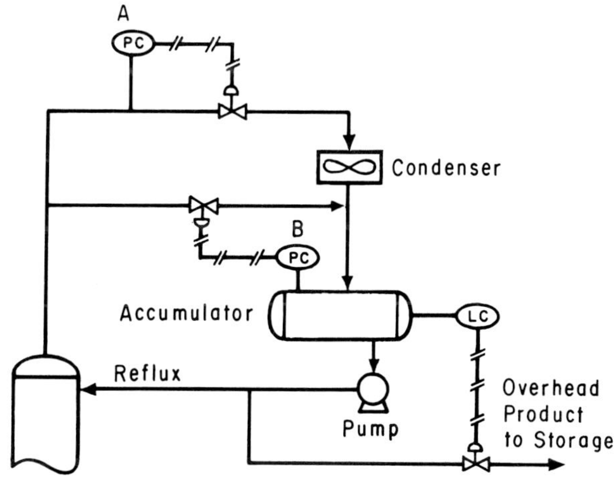

Figure 5.37. Tower Pressure Control with Hot-Gas Bypass

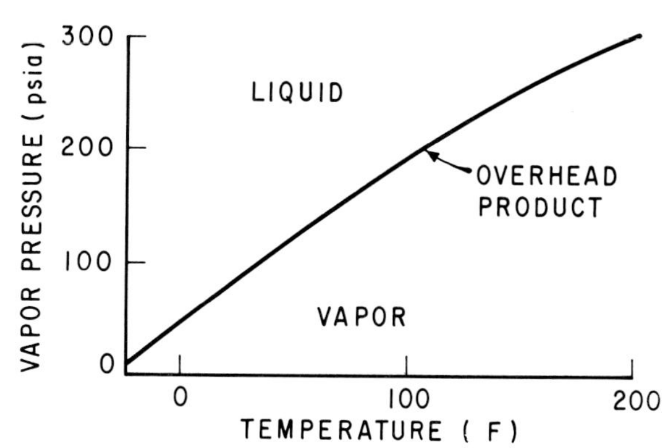

Figure 5.38. Pressure varies according to the condenser outlet temperature.

Figure 5.39 shows a pressure control scheme that may be used when all or part of the overhead product is vapor. Pressure is controlled by regulating the quantity of vapor leaving the accumulator.

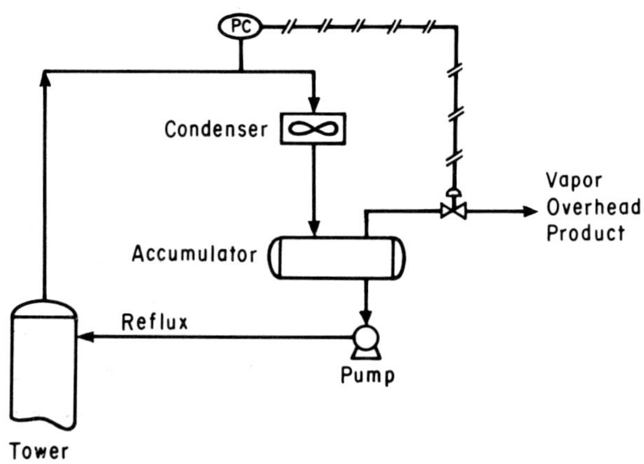

Figure 5.39. Tower Pressure Control by Regulating Flow of Vapor Product

A *flooded-condenser* pressure control is depicted in figure 5.40. Again, this pressure control system uses the vapor pressure versus temperature relation illustrated in figure 5.38. As the tower pressure starts to decline, the control valve on the product line starts to close. This increases the liquid level until part of the condenser tubes is *flooded* with liquid—that is, covered by the liquid. In effect, this reduces the heat transfer surface in the condenser and thereby raises the accumulator temperature. The increase in accumulator temperature raises the product vapor pressure. Therefore, a

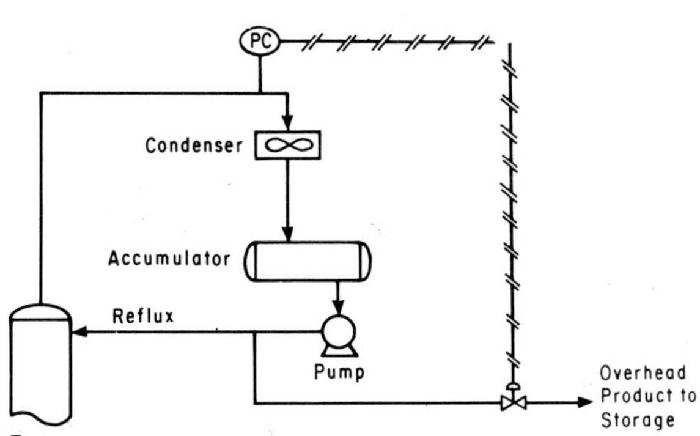

Figure 5.40. Tower Pressure Control by Flooded Condenser

flooded-condenser pressure control actually controls pressure by regulating the amount of condenser heat-transfer surface. One problem that can occur with a flooded-condenser pressure system is that due to noncondensables. If it is assumed that the tower is a depropanizer and the upstream facilities leave too much ethane in the depropanizer feed, then some of the ethane will not condense unless the accumulator temperature is very low. The uncondensed ethane vapor increases the accumulator pressure and thereby increases the tower pressure. The pressure increase causes the product valve to open, and the accumulator may run dry. Controllers similar to figure 5.41 solve this problem. Normally, the pressure controller regulates the product valve, valve *A* on the drawing. However, if

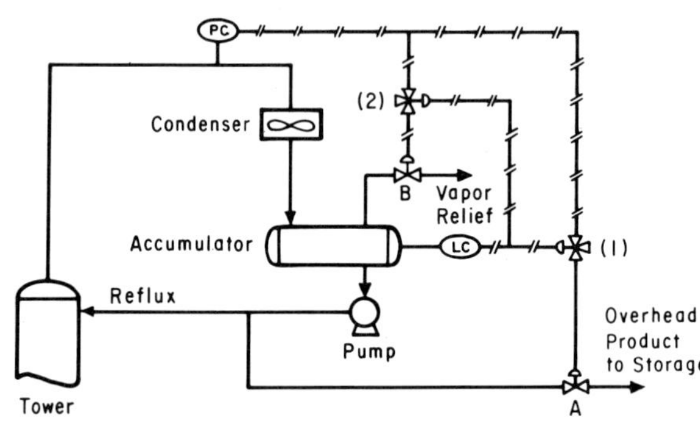

Figure 5.41. Control Equipment for Noncondensables

the liquid level drops too low, the level controller takes control. It causes the three-way valve to vent the air pressure on the product valve diaphragm. This closes the product valve and stops the flow of overhead product to storage. The level controller also causes the three-way valve to stop venting the air pressure on the diaphragm of valve *B*. Now the pressure controller is controlling the tower pressure by regulating the release of ethane vapor from the accumulator. When the liquid level has sufficiently increased, the level controller reverses the actions of the three-way valves. When this is done, the normal operation of the pressure control system will have been restored.

Chapter VI

Cryogenics

Cryogenics is a branch of physics that deals with the production and effects of very low temperatures. In physics, the cryogenic temperature range is usually considered to be below -150 F. However, the natural gas processing industry normally considers cryogenic temperatures to be below -50 F. This is approximately the minimum temperature of a propane refrigeration system.

Cryogenic processing requires the proper combination of pressure and temperature to achieve the desired product recovery. Figure 6.1 shows the proper combination of pressure and temperature for 60 percent ethane recovery from a Gulf Coast natural gas. It can be seen that very low temperatures are required. Processing at these low temperatures requires special techniques.

TYPICAL APPLICATIONS

Expander Process

Figure 6.2 illustrates the flow for an expander process. The inlet gas is first prepared for processing by removing the water and other contaminants that might cause problems. Cooling of the inlet gas is accomplished by using cold residue gas and cold demethanizer internal liquid. The inlet gas is divided so that part flows through the first gas/gas exchanger and is cooled by the residue gas. The remainder of the inlet gas is cooled in the demethanizer side exchanger by withdrawing cold liquid from the demethanizer. Actually, the demethanizer side exchanger serves two purposes: (1) it assists in cooling the inlet gas and (2) it adds part of the heat necessary for properly demethanizing the plant product.

The inlet gas from the first gas/gas exchanger and that from the demethanizer side exchanger join to flow through the second gas/gas exchanger. Liquid condensed from the inlet gas is separated from the vapor in the expander inlet separator. The liquid is fed to the demethanizer. The vapor flows through the expander, where the pressure is reduced, and then to the top of the demethanizer. The enlarged top of the demethanizer serves as the expander outlet separator.

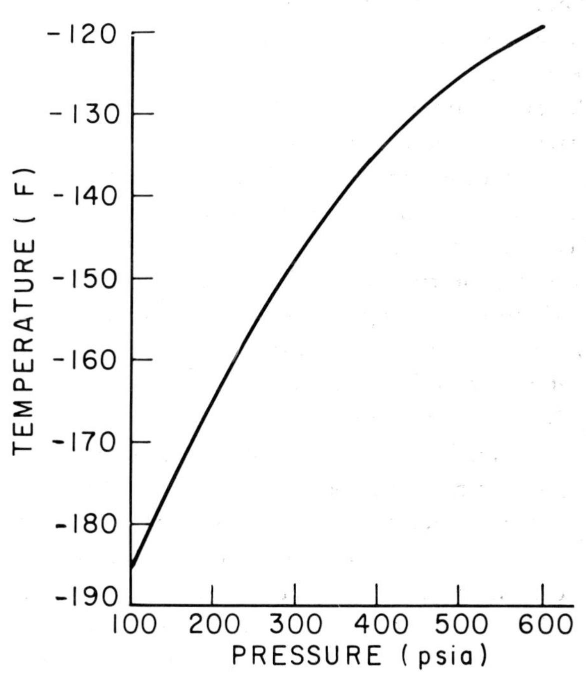

Figure 6.1. Pressure and Temperature to Recover 60 percent Ethane

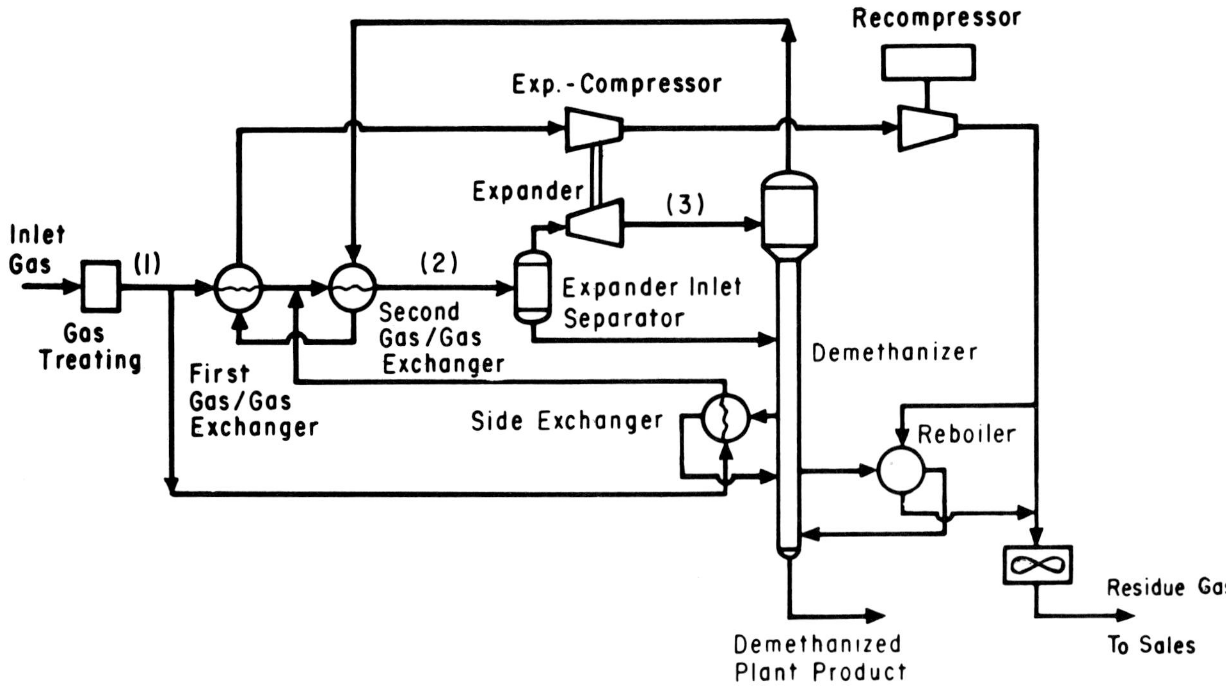

Figure 6.2. Plant Utilizing Expander Process

Cold residue gas from the top of the demethanizer is first used to cool the inlet gas. Then the residue gas is compressed to sales gas pressure by the expander-compressor and recompressor. The expander-compressor is driven by the expander, while the recompressor driver may be a gas engine, gas turbine, steam turbine, or electric motor. Hot residue gas from the discharge of the recompressor is used for heat in the demethanizer reboiler. After final cooling, the residue gas is delivered to sales.

Figure 6.3 represents the pressure-temperature diagram for this expander process. Refer first to the solid line. This represents the plant inlet gas. The solid line on the right is the dew point line. At a fixed pressure, and if the temperature of the gas is to the right of this dew point line, then the gas is 100 percent vapor. If the gas is cooled, liquid starts to condense when the temperature reaches the dew point line. As cooling continues, more liquid is condensed until the bubble point line is reached—the solid line on the left. At this point, all of the gas is liquid. Additional cooling results in colder liquid.

Downstream of the gas treating facilities, the inlet gas is represented by point *1* on both figures 6.2 and 6.3. Look at figure 6.3; as the gas is cooled by the gas/gas exchangers and demethanizer side exchanger, its temperature moves along the dotted line to point *2.* At *2,* the gas enters the expander inlet separator where the condensed liquid is separated from the vapor. This vapor now has its own pressure-temperature diagram, as represented by the dashed curve. At the expander inlet, the gas is on its dew point line.

As the gas flows through the expander, its pressure-temperature path is shown by the dashed line from point *2* to point *3.* Point *3* represents the outlet of the expander. The importance of using the expander as a driver for a compressor can be seen in figure 6.2. If the gas had just been expanded without doing any driver work, the expansion path would be from point *2* to point *4.* (This is called a Joule-Thompson expansion.) Both the outlet pressure and temperature would be higher, resulting in a

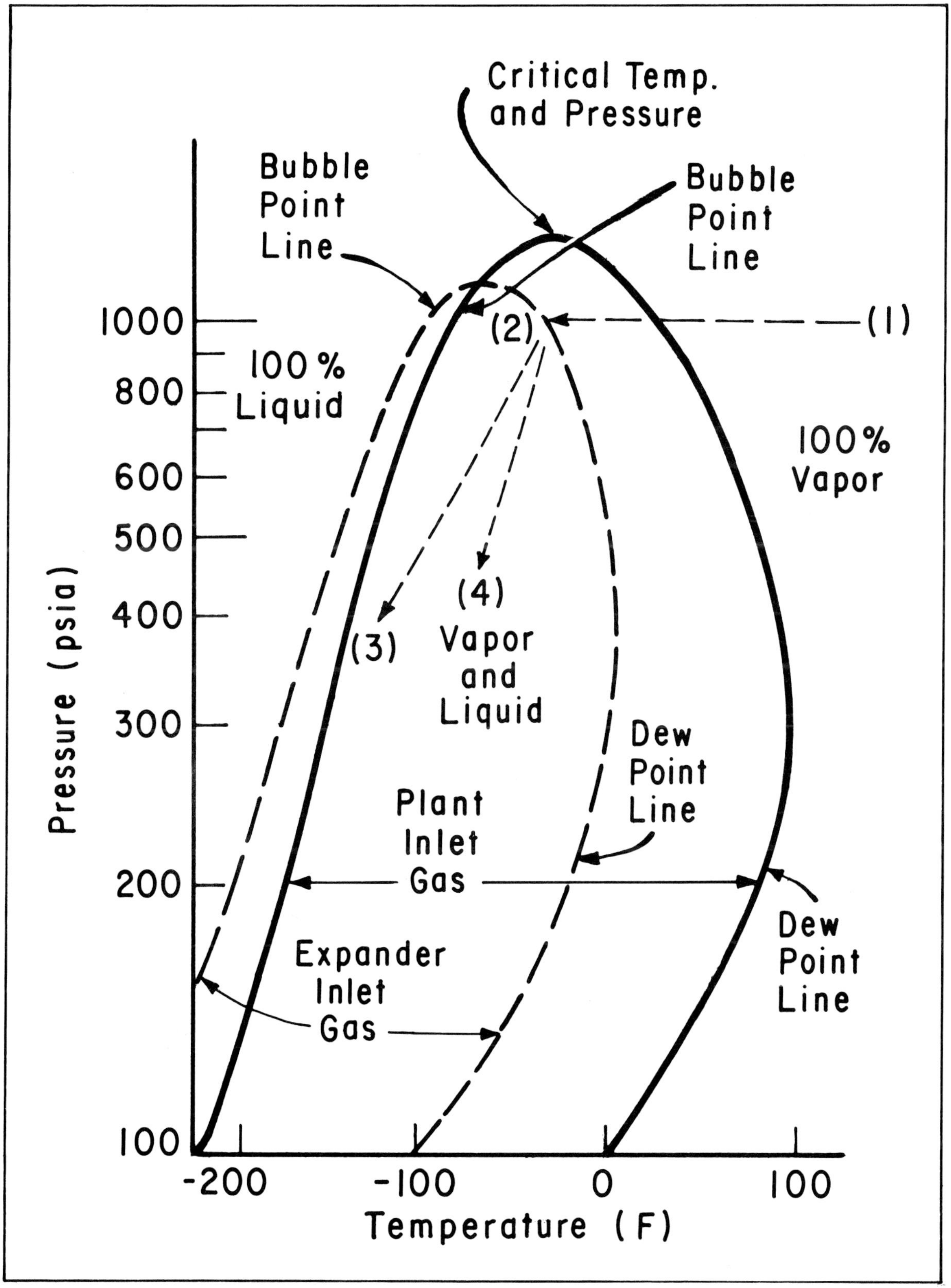

Figure 6.3. Pressure-Temperature Diagram for Expander Process

reduction of product recovery. The outlet pressure is higher because the expander-compressor is not available for compression. A larger compressor would allow a lower outlet pressure, but the outlet temperature would still be higher and thereby reduce product recovery. The principal problems encountered with the expander process are due to (1) gas preparation, (2) product demethanization, and (3) methane condensation.

The gas must be properly prepared for processing. Usually, the most difficult problem is removal of the water. A moisture analyzer can be used to monitor the water content of the gas entering the heat exchanger and related equipment. Surveillance of the inlet gas pressure drop as it flows through the various heat exchangers is also necessary. An increase in the inlet gas pressure drop indicates stoppage of the inlet gas flow path due to either freezing or some foreign matter, such as desiccant particles from the dehydration unit.

Proper product demethanization primarily depends on the demethanizer bottom temperature. The correct bottom temperature depends on the product recovery and demethanizer pressure. Low ethane recovery requires a higher bottom temperature. If the expander-compressor or one of the recompressors is down, then the demethanizer pressure must be increased. A higher bottom temperature is required at a higher operating pressure.

The temperature most important to high product recovery is the vapor temperature to the expander. A lower temperature to the expander increases the product recovery. However, while the expander process condenses the product to be recovered (that is, ethane, propane, butanes, and natural gasoline), it also condenses a considerable quantity of methane. Too much methane condensation can overload the demethanizer. A lower temperature to the expander increases the methane condensation. Therefore, the vapor temperature to the expander should be as low as possible without overloading the demethanizer.

Cascade Refrigeration

A flow diagram for a cascade refrigeration process is depicted in figure 6.4. Only the principal items of equipment are shown. The use of cascade refrigeration is normally limited to processing inlet gas streams at about 500 psig or less. Higher pressures would result in too much methane condensation.

After gas preparation, the inlet gas is cooled by cold residue gas and refrigeration. The cascade refrigeration system is really two refrigeration systems. Since propane can be condensed at a relatively low pressure by air coolers (though water coolers can be used instead of air coolers), then propane is the high-stage refrigerant. Propane is used in the first gas chiller to cool the inlet gas to about -40 F.

Ethane is used as the low-stage refrigerant. The inlet gas can be cooled to approximately -120 F by using ethane in the second gas chiller. Condensing the ethane from the ethane compressor discharge requires a low-temperature coolant. For instance, to condense ethane at 90 F requires a pressure of about 690 psig. By using propane as the ethane condensing coolant, the ethane may be condensed at about 120 psig and -30 F.

Therefore, the propane refrigeration system is used to cool the inlet gas and to condense the ethane refrigerant. Full utilization must be made of the propane refrigerant in the first gas chiller; otherwise, additional inlet gas cooling will be required from the ethane refrigerant in the second gas chiller. Increased ethane refrigeration requires more propane refrigerant for condensing the ethane.

Liquefied Natural Gas (LNG)

There are many means of liquefying natural gas, most of which are, in part, proprietary. Some of the common methods are expander-compression combination, cascade methane-ethylene-propane, mixed refrigerant, multicomponent, and propane-MRC.

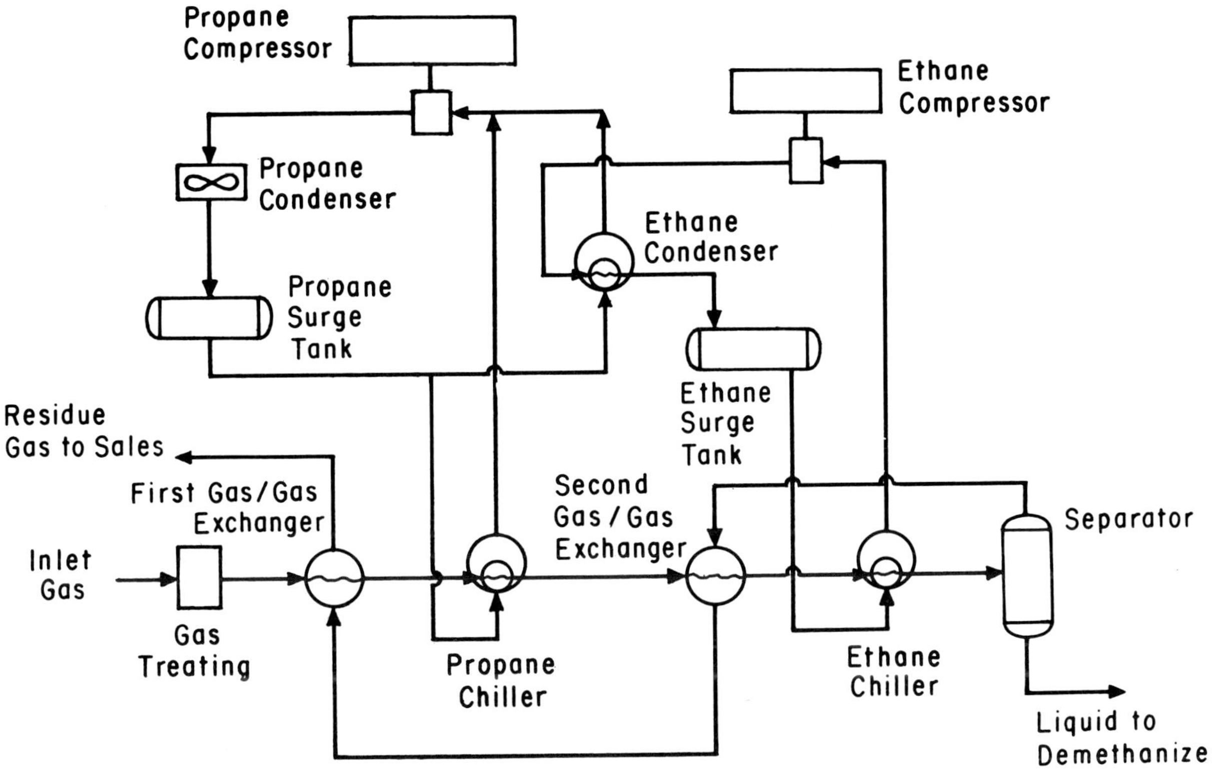

Figure 6.4. Cascade Refrigeration

Figure 6.5 is an LNG cascade cycle. The heat exchangers are of special design, using either a plate fin aluminum core or aluminum tubing wound on an aluminum spool. This type of design requires a large number of different flow streams, which is not easy to obtain with the usual shell-and-tube exchangers. Stream number *1* is the gas to be liquefied. The gas must be dehydrated and all contaminants removed before it enters the plate fin exchangers. Gas stream number *1* is cooled with heavy ends condensing at a point where they can be removed to avoid triple-phase mixtures. The heavy ends are generally removed at approximately -80 F, as shown in stream number *2*. The nitrogen, methane, ethane, and some propane are liquefied at approximately -260 F at about 20 psia and are removed as stream number *3*.

The refrigeration system uses propane as the primary refrigerant, ethylene as a secondary, and methane as the final refrigerant. Ethane can be used as an intermediate refrigerant, but the total horsepower of the refrigeration system will be higher because ethane will not lower the intermediate temperature as low as ethylene, making the methane compression ratio too high for optimum usage of the cascade cycle.

TURBOEXPANDERS

In about 1960, the natural gas and other hydrocarbon industries began using turboexpanders. Since that time, they have had wide and increasing acceptance and application. The sizes found in the hydrocarbon industry usually range from about 250 horsepower up to a few that have reached the 10,000 horsepower range. Currently the average size is approximately 2,000 horsepower.

Turboexpanders are most widely used in the hydrocarbon industry for the recovery of ethane and propane from natural gas. There is also considerable use for them in recovery of power from the expanding streams during the processing of helium, carbon dioxide, and hydrogen. Turboexpanders are also used in the liquefaction of helium, hydrogen, and natural gas.

The heart of the cryogenic process necessitates a machine that is capable of generating low-temperature refrigeration. This requirement is met by the turboexpander. The turboexpander, when used to drive a compressor, has as its fundamental purpose the generation of low-temperature refrigeration. Figure 6.6 shows an installed turboexpander-compressor. From the curve in figure 6.7, it can be noted that the relative efficiency of the turboexpander is competitive with, and even better than, mechanical compressor drives at temperatures below about -75 F. The turboexpander is in-

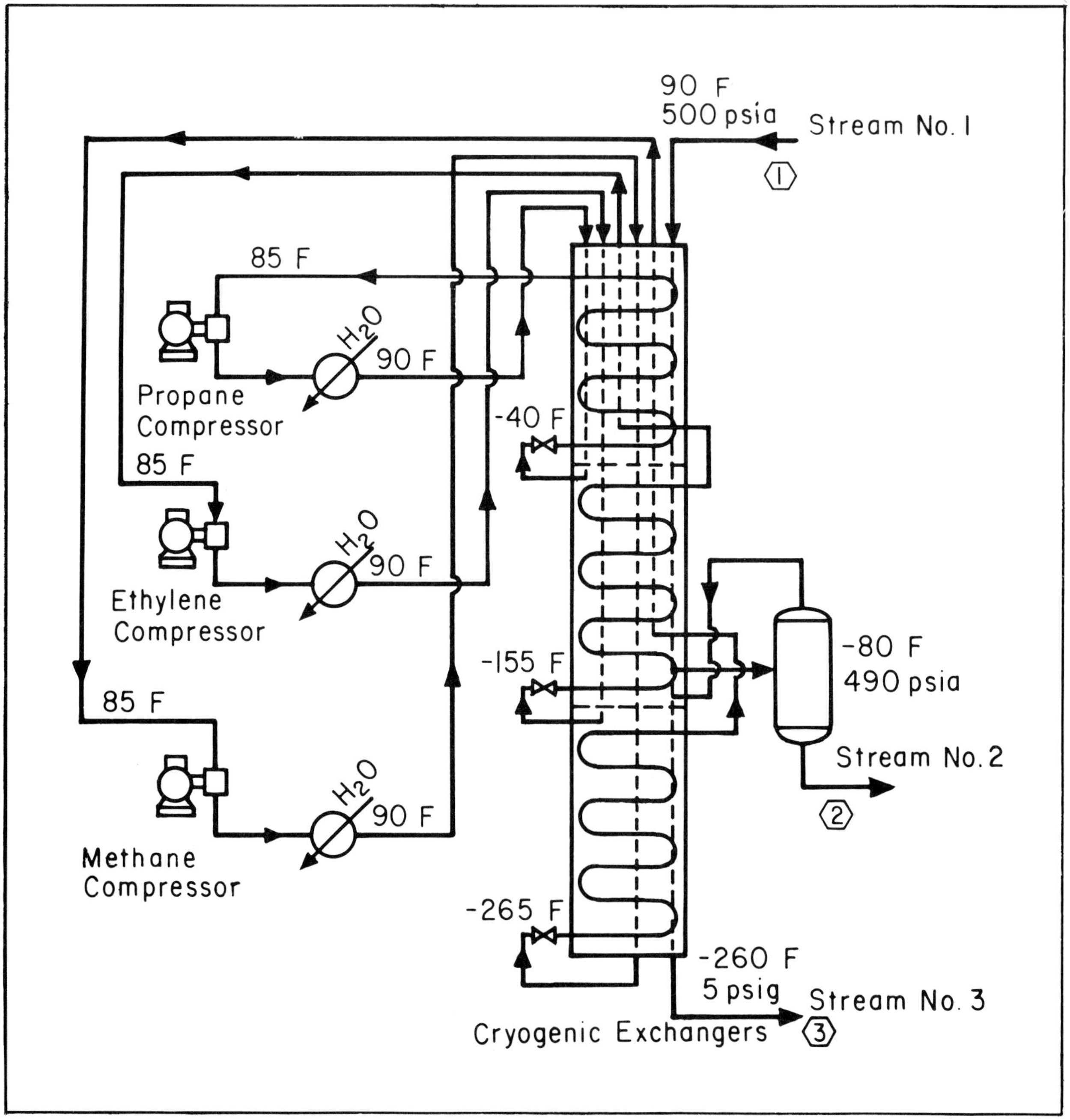

Figure 6.5. LNG Cascade Cycle

Figure 6.6. Turboexpander-Compressor

timately tied in with the system in which it is applied, and each is sensitive to the characteristics of the other.

A Btu subtracted by refrigeration, which is the opposite of heating, at low temperatures is sevenfold more valuable in terms of work required than a Btu of heat added in most petroleum processes. For this reason, refrigeration processes must be more completely and carefully insulated to prevent unwanted heat gains than would be justified to prevent heat losses at the higher temperatures used in most petroleum processes. The turboexpander should be located in the system at the lowest temperature position possible in view of the efficiency-temperature relationship shown in figure 6.7.

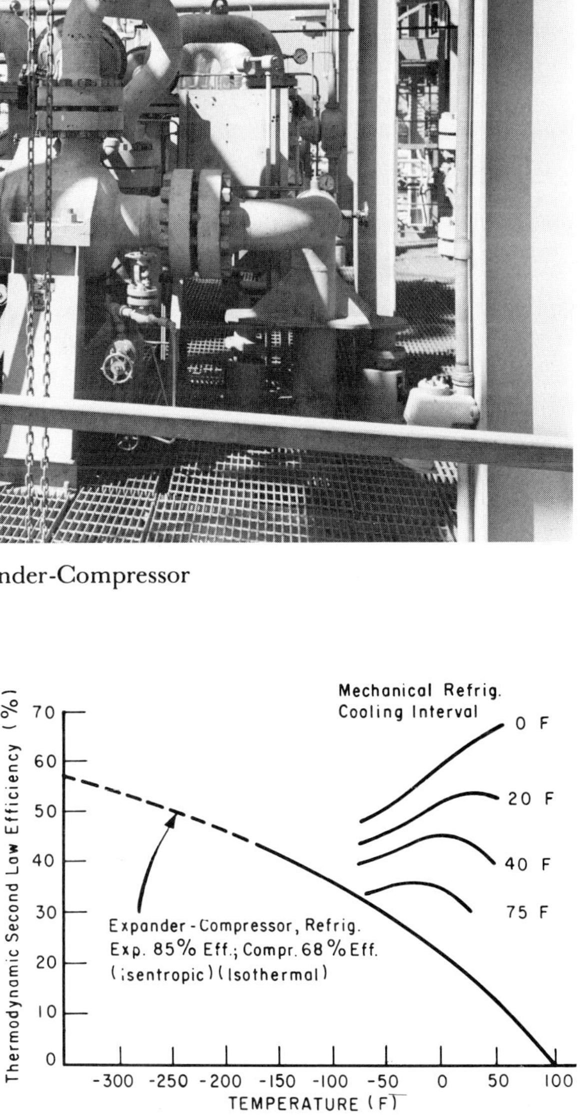

Figure 6.7. Mechanical versus Turboexpander Refrigeration Efficiencies

There is also the consideration of the control of the turboexpander and the absorption of the turboexpander power in the planning of the system. It simplifies the control if the power from the turboexpander can be absorbed by a compressor in which the mass flow is proportional to the turboexpander flow. Then the power is absorbed at all flows at speeds that rather closely approach the speeds for optimum efficiency. This was illustrated in the flow diagram of figure 6.2. With this arrangement, speed control is not normally required because the speed adjusts itself within a reasonable range due to the power balance between the compressor and the expander.

In most applications, the turboexpander has means for controlling the flow in a manner that reasonably well utilizes the expansion energy and does not waste it in throttling. The older flow-control method utilized primary nozzles divided into several groups, each with a valve, so that one or more valves could be opened to select the rate of flow. This is applicable only to impulse turbines, but they operate at lower efficiency than reaction-type turbines and are no longer widely used.

The radial reaction turbine attains the highest efficiency. The variable nozzle assembly surrounds the rotor. The nozzle blades can be tilted to vary the clearance between them so that the available flow area is varied in the nozzles that inject the gas into the rotor. This is effective in controlling the flow in a turboexpander over a wide range, even above design flow. The fixed losses in the turboexpander rapidly reduce overall efficiency at low flows because these losses must be prorated against a lower flow. A combination of rotor and nozzle that gives a high efficiency over a wide range is available.

To operate the turboexpander at the lowest possible temperature to attain the best process efficiency, it usually must operate on a gas stream that partially condenses as it passes through the turboexpander, and this places special requirements on the turboexpander. Turboexpanders that claim no efficiency loss for any amount of liquid condensed are available.

In addition to being capable of wide flow variation, some applications require rather wide pressure variation. When this is a requirement, it is best to have a loading device consisting of a compressor or a turbulence device, the speed of which responds to some degree to the applied power. Then the speed will closely follow the speed corresponding to optimum efficiency.

The turboexpander is a small part of the investment in a cryogenic plant. Yet, its performance in terms of efficiency and reliability are important to the success and profitability of the process. The design, start-up, and operation of each installation must carefully avoid exposing the turboexpander to severe operating conditions. Otherwise, the plant may shut down or have its performance impaired.

In most applications, the turboexpander is started and operated for a while at reduced speed for various reasons (such as cooling down the plant or bringing other units on stream) so that it may not reach full speed for quite some time. If the turboexpander is to be capable of operation at various speeds, it cannot be of flexible shaft design but rather must be of stiff shaft design to avoid shaft critical speed. This also applies to the bearings that are used. They must not be of a design in which the oil film is critical within the operating speed. The bearings should be designed to carry heavy loads created by the imbalance that is caused by freeze-up or erosion. Such heavy-duty bearings have the added quality of avoiding oil-film resonance mentioned above. These bearings require the shaft to be of large diameter, which, in combination with rigid bearings, maintains excellent seal alignment. Some turboexpanders are built with tilting shoe bearings. The purpose of tilting shoe bearings is to suppress oil swirl. This phenomenon occurs at roughly twice the frequency of the oil-film resonance, so if oil-film resonance is out of the way, then oil swirl is certainly not a problem.

Another potentially serious problem in turboexpanders is the possibility of resonance in the rotor that is excited by a frequency equal to the running speed multiplied by the number of blades in the rotor—a frequency on the order of several thousand cycles per second. If such resonance is encountered in operation, it will destroy the rotor in a short time; thus, the rotor must be designed so as to have the operating range free of all such resonance frequencies.

SPECIAL TECHNIQUES

Gas Preparation

The first consideration in cryogenic processing of natural gas is contamination. Contaminants could be considered to be water, carbon dioxide, hydrogen sulfide, nitrogen, normal butane, normal pentane (and all of the hydrocarbons of six or more carbon atoms), dirt, iron sulfide, crude oil, salt water, wax, or lubricating oil. CO_2, H_2S, H_2O, nC_4, nC_5, and C_6+ form triple-phase mixtures under some temperature-pressure relationships. There has been very little experimental work on the solubility of these components in the liquid phase. The solubilities of these contaminants vary with temperature, pressure, and composition. The best way to avoid trouble is to remove contaminants if temperatures equal to or below their pure component, solid temperature-pressure relationship are anticipated. Still better, remove them from the system in all cryogenic processing, unless their solubility limits are quite apparent.

Heavy hydrocarbon contaminants can be removed at the temperature-pressure relation at which they become liquefied or where economics justify. Salt water, wax, dirt, scale, iron sulfide, and oil (both lubricating as well as heavy ends) carried over from wellhead separators and compressors can be removed by filtering or electrostatic precipitation. These contaminants cause exchanger fouling and triple-phase composition that results in solids buildup, which, in turn, results in flow stoppage. Cryogenic exchangers are delicate and have small passages that are not easy to clean. A speck of dirt on an exchanger surface causes the growth of solid CO_2, water, or other contaminants. These solids plug flow passages and require a warmup to rid the system of them.

H_2S and CO_2 removal are handled with the usual processes, but the selection of the treating process should consider the cryogenic level of processing. Deep cryogenics (very low temperatures) such as required for LNG require the almost complete removal of water by a molecular sieve. A treating process should be chosen that is compatible with a molecular sieve. Keep in mind that the treating process can be a source of contamination itself.

Cryogenic Heat Exchanger

Cryogenic gas processing requires specialized types of heat exchangers. Below -50 F, the common shell-and-tube heat exchanger becomes uneconomical in comparison to the specialized class of exchangers known as plate fin, or core exchangers. Most cryogenic exchangers are made of aluminum because of its increasing tensile strength as the temperature drops. From a cryogenic standpoint, aluminum has two poor qualities: very large thermal expansion and fatigue cracking from repeated stresses. A core exchanger is rated by the cubic foot of core and is built as a rectangular solid with external headers. The design lends itself to a wide variety of flow streams with a large extended surface per cubic foot of core. The core is constructed of an aluminum plate, an aluminum corrugated section, a second aluminum plate, a second corrugated section, and so on until the desired core is built up. The corrugated areas can be either parallel flow or cross flow. A great variety of exchangers can be built up by a proper selection of headers and side bars to seal the core where desired. This type of exchanger is very compact and has very high performance. The exchangers are rated with 2 F to 3 F approach temperature and very little pressure drop. The heat transfer with this type of exchanger is very efficient, and a temperature cross can be handled with ease.

Construction Material

All materials whose tensile strength increases with decreased temperatures are excellent for cryogenic gas processing application. There is a large variety of grades of aluminum suitable for cryogenic construction, all of which have very large thermal-expansion properties. There is a great variety of nickel steels available for cryogenic construction. Nickel steels have lower thermal-expansion properties than aluminum; the higher the nickel content, the lower the thermal expansion of the material.

Tensile, yield, and fatigue strengths, brittle fracture, notch toughness, thermal conductivity, and specific heat of materials take on a greater significance at the low temperature for cryogenic applications than would apply at normal temperatures. Codes require the design of equipment at ambient temperatures, but at cryogenic temperatures, the safety factor is much greater than the usual design safety factors.

Pumps

There is a unique class of pumps available for cryogenic processing. Aside from the type of construction materials, which are the same as for core exchangers, they employ extended shafts between the motors and pump bodies. This type of construction allows these pumps to operate deep inside the insulation of the cold box, with the motor external to the pump's insulated housing.

Instruments and Control Valves

Control valves use extended shafts so that the body can be operated in the cold box with the top works external of the insulation. Block valves use the same type of construction. Instruments are mounted external to the cold box with internal connections.

Physical Data

There is a vast amount of research being conducted at universities and by private industry for the measurement of multicomponent equilibrium and enthalpy (heat content) data. Computer programs have been written to utilize this data. GPA, AGA, IGT, and others have sponsored much of the research. Equilibrium data depends on temperature, pressure, and composition of the hydrocarbons. The process engineer must know the valid area of the equilibrium data he is using. The GPA equilibrium ratios are a good example of the composition dependency of *K values,* where the K value is the ratio of the molecular fraction of a component in the vapor phase to the molecular fraction of that component in the liquid phase. The very nature of cryogenic gas processing involves phase behavior and lends itself to dew point, bubble point, flash calculations, and heat balance. These calculations are of the "hunt and try" variety, in which there are one or more unknowns in the equations describing the system. One must assume a value for the unknowns and verify them by equilibrium calculations.

Cold Box

A cold box is a common insulation around the cryogenic process equipment to minimize the heat loss. The equipment is usually located in the cold box with the coldest operating equipment in the center and the warmer near the outside. The cold box consists of a structural frame to support the outside cover around the process equipment. The insulation is installed around the equipment, but it is confined by the outside cover. Nitrogen gas is released inside the cold box to avoid explosive mixtures from gas leaks. The insulation is usually expanded perlite, fiberglass, or foam plastic, depending on temperature level. Sample connections are brought to a common place at the bottom of the cold box. The sample lines are usually stainless steel or aluminum tubes with stainless-steel valves.

CALCULATIONS

Expanders

The pressure level of the expansion is usually set by other process requirements so that generally the outlet temperature and vapor-liquid composition are to be determined at this pressure. If access to a computer that is capable of a flash calculation is possible, the task is simple. Manual calculations are more difficult. The calculation can be accomplished by first determining the enthalpy and entropy of the gas composition at the temperature and pressure of the expander inlet. This is shown graphically in figure 6.8; the expansion starts at point *A*. The next step is to calculate a flash at the desired expander outlet pressure (point *B*) at several assumed temperatures. The entropy (randomness of the mixture) of the resulting vapor-liquid mixture is calculated and matched with the entropy of the expander inlet, and then the flash temperature is chosen that gives the best entropy match—that is, $S_1 = S_2$ at point *B*. The difference in enthalpy ($h_1 - h_2$) of the inlet and outlet of the expander at 100 percent efficiency is calculated. The enthalpy difference is multiplied by the expander efficiency—that is, ($h_1 - h_2$) 77% = $h_1 - h_3$. Then the new expander outlet enthalpy (h_3) is determined by using this new enthalpy difference. The inlet gas is flashed at the expected outlet pressure at several temperatures until there is an enthalpy match of the resulting vapor-liquid with the other enthalpy determined above. The shaft horsepower of the expander is this enthalpy difference divided by the mechanical equivalent of heat (2,545 Btu=1 hp), times the pounds per hour of flow through the expander, times the mechanical efficiency of the expander. The expander outlet temperature, vapor composition, liquid composition, and horsepower output of the expander have thus been determined.

$$\text{hp} = \frac{(h_1 - h_2) \times \text{effic.} \times \#/\text{hr}}{2{,}545}$$

$$= \frac{(h_1 - h_3)\ \#/\text{hr}}{2{,}545}$$

Refrigeration Systems

1. Determine the refrigeration load (MBtu/hr) from plant process. Normally add 5 percent for losses through insulation.
2. Determine refrigeration level (temperature). Normally 10 F below minimum process temperature.
3. Select the refrigerant and determine the refrigerant condensing temperature and pressure. For aerial condensing, the normal temperature is 120 F to 140 F. For water condensing, the normal temperature is 90 F to 100 F.
4. Determine the number of compressor stages required and the type of refrigeration cycle. Normally one stage is used on propane refrigerant temperatures above 30 F. Two stages are used between 30 F and -40 F, and cascade systems are used below -40 F.
5. If two or more stages are used, determine the interstage pressure and temperature. $P_2 = P_1 \cdot P_3$ for two stage where P_1 is evaporator pressure, P_2 is interstage pressure, and P_3 is second-stage discharge pressure.
6. From the Mollier chart or other enthalpy data, determine the pounds per hour of refrigerant required.

   $$W_1 = \frac{\text{Duty}}{h_4 - h_2},$$

   where W_1 is pounds per hour of refrigerant, h_4 is enthalpy in Btu per pound (Btu/#) of refrigerant vapor at evaporator pressure P_1, and h_2 is enthalpy (Btu/#) of refrigerant to evaporator.
7. If an economizer flash tank is used, determine the flash vapor in which flash vapor

   $$W_2 = \frac{(W_1)(h_2 - h_1)}{(h_1 - h_3)},$$

   where W_1 is liquid to chiller, h_2 is enthalpy (Btu/#) of liquid to chiller, h_1 is enthalpy (Btu/#) of liquid from surge tank, and h_3 is enthalpy (Btu/#) of vapor from flash tank.
8. Determine the compressor discharge temperature and horsepower. This can be done either by conventional horsepower calcula-

tions, called the *N* method, where

$$hp = ZRT_{1}\frac{n}{(n-1)}\left\{\left(\frac{P_2}{P_1}\right)^{\frac{n-1}{n}} - 1\right\}$$

or by the Mollier diagram method. For pure gas (only one component), the Mollier method is preferred.

9. Correct for adiabatic efficiency. For centrifugal compressors, the adiabatic efficiency (Nad) is normally between 65 percent and 75 percent. For reciprocating compressors, the adiabatic efficiency is between 75 percent and 85 percent.
10. For the Mollier chart method, find the suction pressure, temperature, and enthalpy on the Mollier chart. Move up the chart at constant entropy (isentropic) to the discharge pressure and read the enthalpy of the gas. To find the actual discharge enthalpy,

$$h_d = \frac{(h_2)ad - h_s}{Nad} + h_s,$$

where h_d is actual discharge enthalpy, $(h_2)ad$ is discharge enthalpy at constant entropy, h_s is the suction enthalpy, and Nad is adiabatic efficiency.

11. Find the discharge temperature from the Mollier chart.
12. Determine the compressor horsepower where W=#/ hr gas.

$$BHP = \frac{(W)(h_d - h_s)}{2{,}545}$$

13. If two stages are used, find the second-stage suction temperature, pressure, and enthalpy and proceed as in step 10.
14. Determine the refrigerant condenser duty.

$$Duty = W(h_7 - h_1),$$

where W is #/hr refrigerant to condenser, h_7 is enthalpy of gas to condenser, and h_1 is enthalpy of liquid from condenser.

Sample Problem

1. Assume 1,000 tons of refrigeration, 12,000 MBtu/hr, are required for a process.
2. The process streams are to be chilled to -20 F. The evaporator temperature will be -30 F.
3. Use propane as a refrigerant. Water condensing will be used with the propane condensing at 110 F.
4. Use two stages of refrigeration with a flash tank economizer as shown in figure 6.9.
5. From the Mollier chart for propane (fig. 6.10), the condensing pressure at 110 F would be 215 psia. Allowing 5 psi pressure drop through the condenser would require a 220 psia second-stage discharge pressure. The evaporator pressure or first-stage suction pressure would be 20 psia at -30 F. The interstage pressure will be

$$\sqrt{\frac{220}{20}} \times 20 = 67 \text{ psia.}$$

6. Find refrigerant to chiller. (See Mollier chart fig. 6.10.)

$$W = \frac{Duty}{h_4 - h_2} = \frac{12{,}000{,}000}{101 - (-44)} = 82{,}759 \ \#/hr$$

7. Find flash tank vapors.

$$W_2 = (W_1)\frac{(h_2 - h_1)}{(h_1 - h_3)} = 82{,}759\frac{(-44 - 6)}{(6 - 118)}$$

$$= 36{,}946 \ \#/hr$$

8. Find first-stage horsepower. On Mollier chart find suction pressure 20 psia and enthalpy 101 Btu/#(h_4). Draw a line at constant entropy to 67 psia and find h'_5=128 Btu/#.
9. Use reciprocating compressors and assume the adiabatic efficiency is 75 percent.
10. Calculate discharge enthalpy.

$$h_5 = h_4 + \frac{h'_5 - h_4}{0.75} = \frac{128 - 101}{0.75} + 101$$

$$= 137 \text{ Btu/\#}$$

11. From Mollier chart t_5=78 F.
12. First-stage horsepower is

$$\frac{W_1(h_5 - h_4)}{2{,}545} = \frac{82{,}759(137 - 101)}{2{,}545}$$

$$= 1{,}171 \text{ BHP}$$

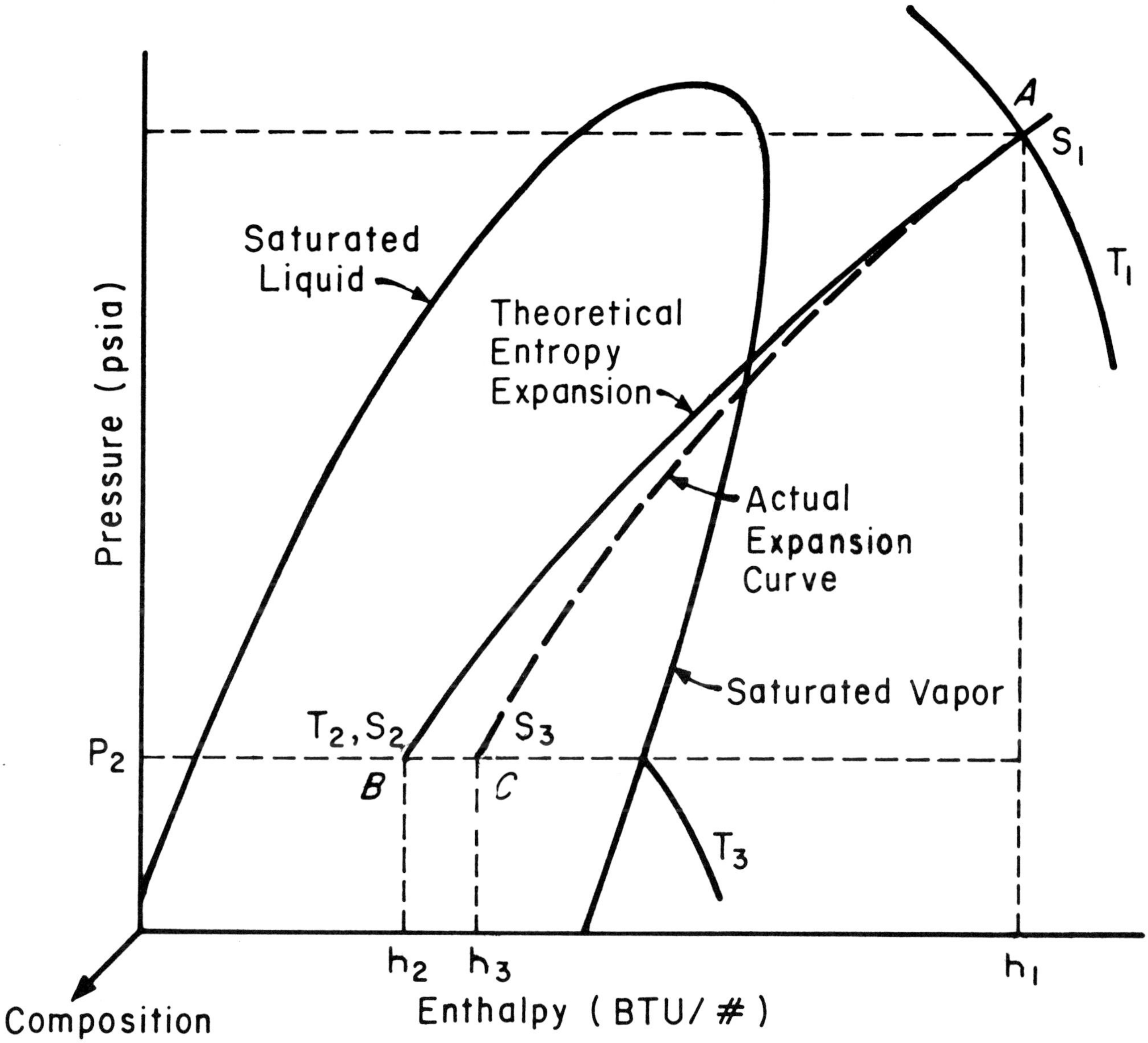

T = Temperature, F
S = Entropy, BTU/#R
h = Enthalpy, BTU/#
P = Pressure, psia

$$HP = \frac{(h_1 - h_2)\,\text{Efficiency} \times \text{Mass Flow}}{2545} = \frac{(h_1 - h_3)\,\text{Mass Flow}}{2545}$$

Figure 6.8. Mollier Chart for Expander Calculations

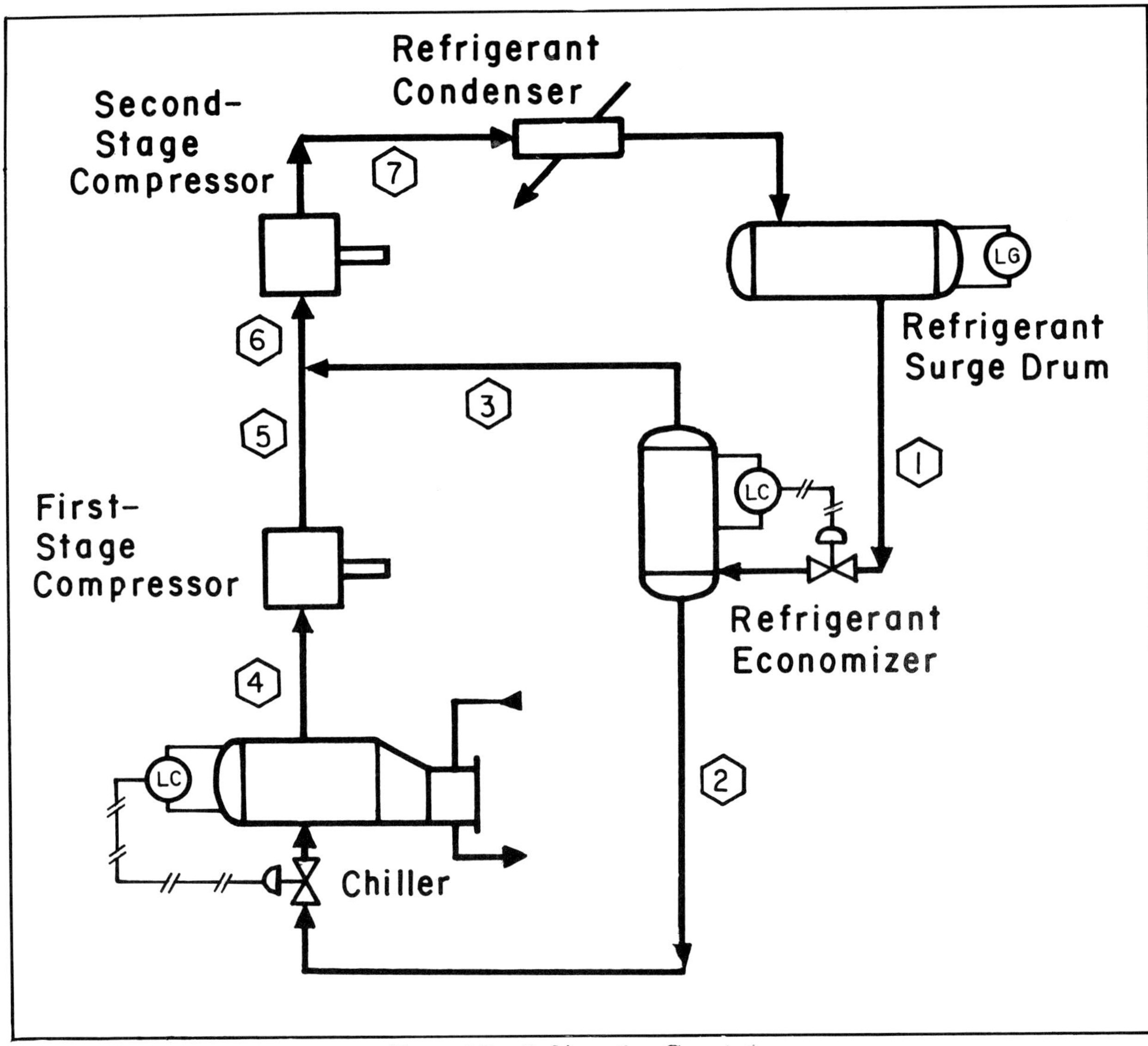

Figure 6.9. Refrigeration Calculations

13. Find second-stage suction conditions.

$$h_6 = \frac{(W_1)(h_5) + (W_2)(h_3)}{W_1 + W_2}$$

$$= \frac{(82{,}759)(137) + (36{,}946)(118)}{82{,}759 + 36{,}946}$$

$$= 131.13 \text{ Btu/\#}$$

$t_6 = 62$ F @ 67 psia $\qquad h'_7 = 157$ Btu/#

$$h_7 = \frac{157 - 131.13}{0.75} + 131.13$$

$$= 165.62 \text{ Btu/\# @ 169 F and 220 psia}$$

Second-stage horsepower

$$= \frac{(82{,}759 + 36{,}946)(165.62 - 131.13)}{2{,}545}$$

$$= 1{,}622 \text{ BHP}$$

14. Find condenser duty.

$$(82{,}759 + 36{,}946)(165.62 - 6)$$

$$= (119{,}705)(159.62) = 19{,}108 \text{ MBtu/hr}$$

Figure 6.11 summarizes the results of the sample problem.

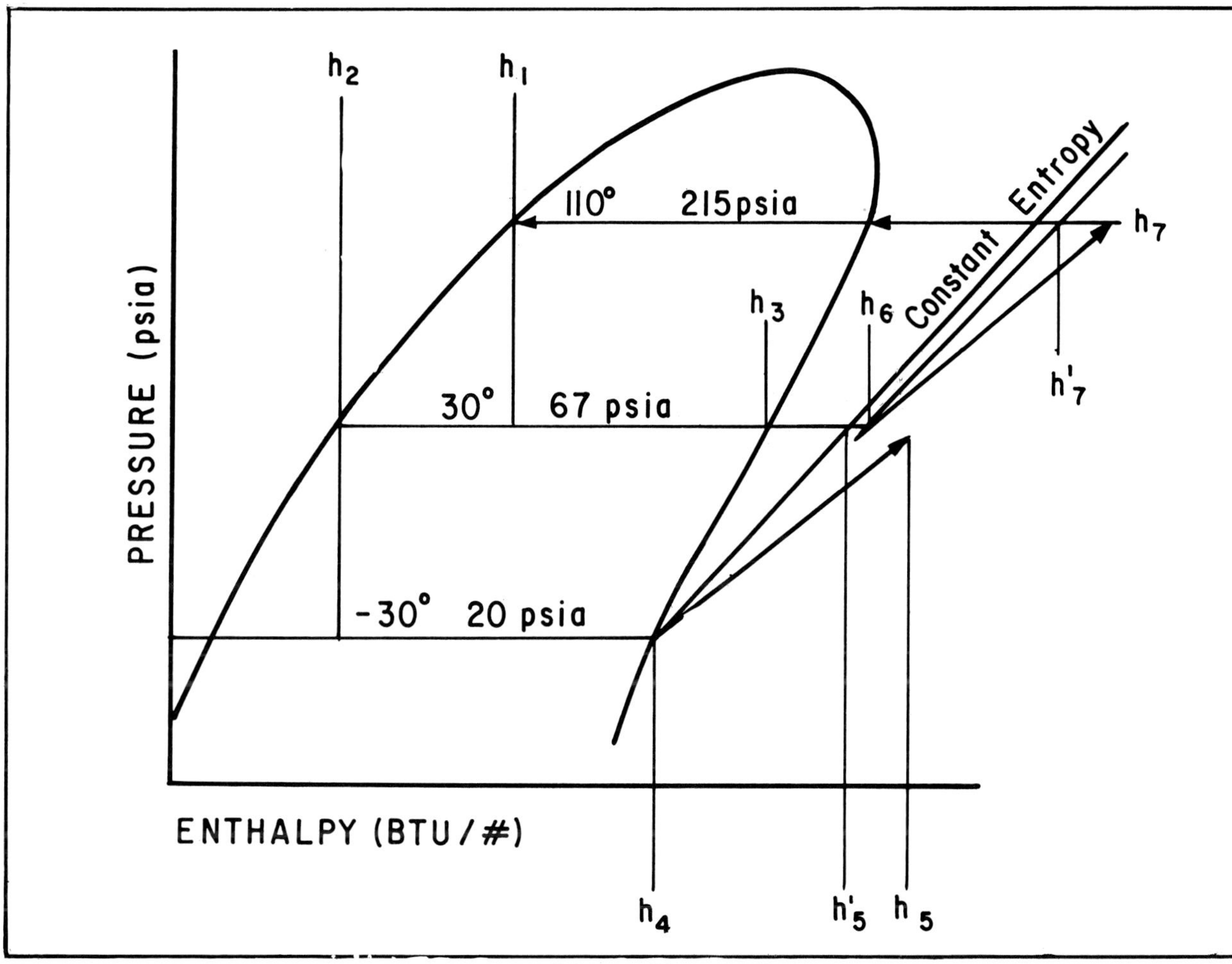

Figure 6.10. Mollier Chart for Propane

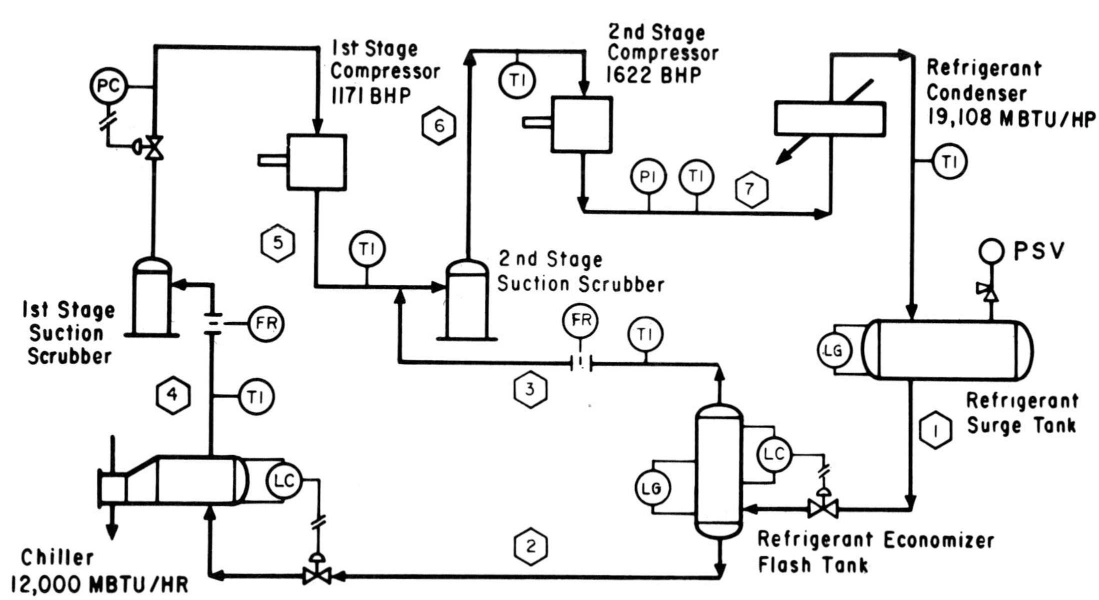

Figure 6.11. Results of Refrigeration Sample Problem

Chapter VII

Gas and Product Treating

Many natural gases must be treated. For this chapter, the term *gas treating* is limited to the removal of sulfur compounds and carbon dioxide (CO_2) from natural gas. The sulfur compound found in largest quantities in natural gas is hydrogen sulfide (H_2S). Both H_2S and CO_2 are called *acid gases. Sweetening* is another name for the removal of sulfur compounds. Other sulfur compounds found in natural gas are mercaptans (RHS), carbonyl sulfide (COS), and carbon disulfide (CS_2). *Sour gases* are those having high concentrations of sulfur compounds. *Sweet gases* are those having small concentrations of sulfur compounds. A common specification calls for less than 0.25 grains of H_2S per 100 standard cubic feet.

Natural gas is frequently treated prior to sale or entry to a process plant. Carbon dioxide has no heating value, and sometimes it must be removed to raise the Btu content of the gas to an acceptable level. Also, it is corrosive in the presence of free water. Sulfur compounds found in gas are corrosive under many conditions, particularly if water or oxygen are present. If they are not removed, they may contaminate liquid hydrocarbon products extracted from the gas. Furthermore, these sulfur compounds are toxic. When they are burned, they form sulfur dioxide (SO_2), a choking, irritating gas.

Gas treating processes can be classified in two ways: (1) *absorption* and (2) *adsorption.* In absorption processes, the gas stream contacts a liquid that selectively removes acid gases. There are many processes in which the gases are physically dissolved in the liquid. The liquid is then regenerated by reducing system pressure to allow the absorbed acid gases to flash out of the liquid. In other absorption processes, the acid gases react chemically with the liquid. These usually require heat to regenerate the liquid and drive off the gases.

Adsorption processes are generally called *dry-bed* units. The sour-gas stream contacts a solid adsorbent that selectively removes the sulfur compounds and/or carbon dioxide. The adsorbent material is regenerated with heat or by lowering the pressure.

A list of most commercial gas treating processes is shown in table 3. More information may be found in the references listed in the bibliography.

AMINE SWEETENING PROCESS

There are a great number of absorption treating processes, but the most common is the amine process. The liquid absorbent is a mixture of water and a chemical amine, usually monoethanolamine (MEA). Most of the following discussion on plant operation is also applicable to other liquid processes.

Process Description

The basic amine flow diagram is shown in figure 7.1. Sour gas enters the bottom of the absorber (contactor) and contacts the lean amine solution, which enters at the top. The amine solution removes the acid gas components, and

Table 3. Some Gas Treating Processes

Process	Numbered References	Gases Removed CO_2	H_2S	RHS	COS	CS_2
Monoethanolamine (MEA)	(18, 26, 27)	X	X		X*	X
Diethanolamine (DEA)	(18, 26, 27, 29)	X	X		X	X
Diglycolamine (DGA)	(9, 18, 27)	X	X		X	X
Sulfinol	(12, 18, 27, 30)	X	X	X	X	X
Hot Potassium Carbonate	(2, 14, 18, 26, 27)	X	X		X	X
Glycol-Amine	(27)	X	X			
Iron Sponge	(26, 27)		X			
Selexol	(1, 18, 22, 30)	X	X	X	X	X
Stretford	(19, 25)		X			
SNPA-DEA	(8, 32)	X	X	X	X	X
Fluor Solvent Process	(4, 18, 26, 30)	X	X	X	X	X
Molecular Sieves	(7, 17, 19, 21, 26)	X	X	X	X	X
Purisol	(24, 26, 30)	X	X			
Catacarb	(18)	X	X		X	X
Benfield	(18)	X	X		X	
Giammarco-Vetrocoke	(5, 18, 19)	X	X			
Rectisol	(18, 30)		X			
Estasolvan	(15, 18, 30)		X	X	X	X
Takahax	(19, 30)		X			
Thylox	(19)	X	X			
Ferrox	(19)	X	X			
Manchester	(19)	X	X			
Perox	(19)		X			
Lacy-Keller	(19, 30)		X	X		
Townsend	(19, 30)		X			
IFP	(19)		X			

*MEA reacts nonreversibly with COS and therefore should not be used to treat gases with a large concentration of COS.

sweet gas flows from the top of the absorber to a scrubber. The absorber may be either a trayed tower or a packed tower.

The rich amine leaves the bottom of the tower and flows through a flash tank. The flash tank vapors are sent to the fuel system and are usually sour. From the flash tank, the amine stream goes through heat exchangers where the stream temperature is raised to approximately 190 F. The stream then flows to the top section of the regeneration stripper—that is, the still. The still bottoms are heated in the reboiler. The heat converts some of the water in the amine solution to steam. This steam rises through the still and heats the counterflowing amine solution. This heat releases the acid gases, and the steam carries them to the top of the still. The steam and acid gases then enter the reflux condenser where the steam is condensed, separated, and pumped back to the still. The acid gases leave the separator and go either to a flare stack or a sulfur recovery plant. The hot lean amine leaves the bottom of the still, is filtered, and flows through a heat exchanger and then to the surge tank. From there, the cooled amine solution is pumped to the top of the contactor.

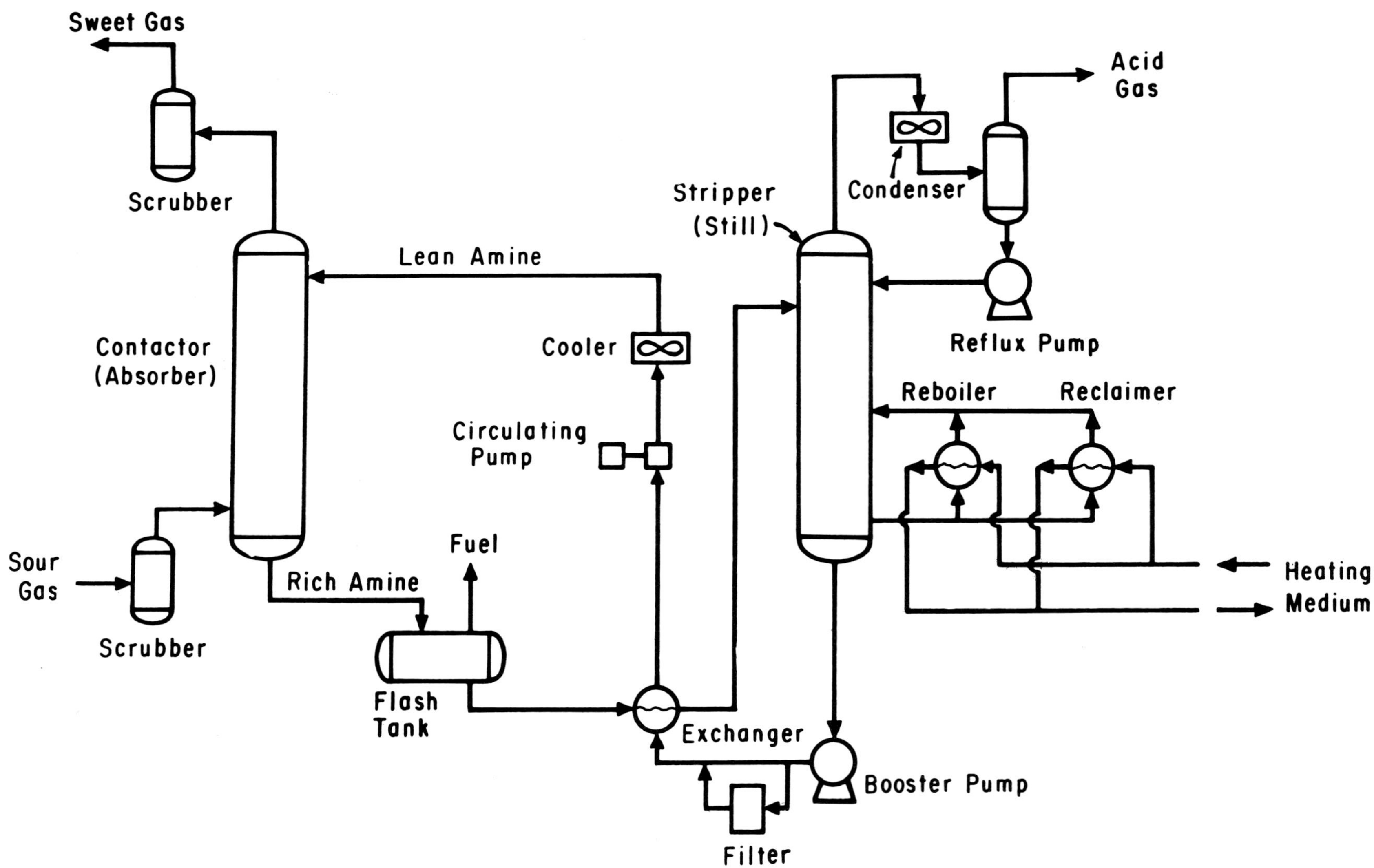

Figure 7.1. Typical Amine Process

A reclaimer is often omitted from an amine plant even though it can greatly reduce operating problems. A flow diagram of a reclaimer setup is shown in figure 7.2. A small portion of the circulating solution (about 1 percent) is run through the reclaimer where it is distilled and returned to the main solution. This removes degradation products, heat-stable salts, and solids. The amine can be released from the heat-stable salts by the slow addition of about 0.03 percent by weight of sodium carbonate to the reclaimer. The maximum temperature of the amine solution should be limited to 290 F. At the end of the reclaimer cycle, the accumulated products in the reclaimer bottoms are dumped.

Operating Amine Plants

Operating problems in amine plants center around five major areas: (1) residue-gas specification, (2) corrosion, (3) foaming, (4) amine consumption, and (5) winterization. Although this section discusses MEA systems specifically, many of these same approaches can be used on other treating processes.

Residue Specifications. If the residue gas is too sour, the inlet gas must be contacted with more *lean* amine. This may be done in the following ways.

1. Amine regeneration can be improved by exchanging more heat in the reboiler to increase the steam stripping rate that will

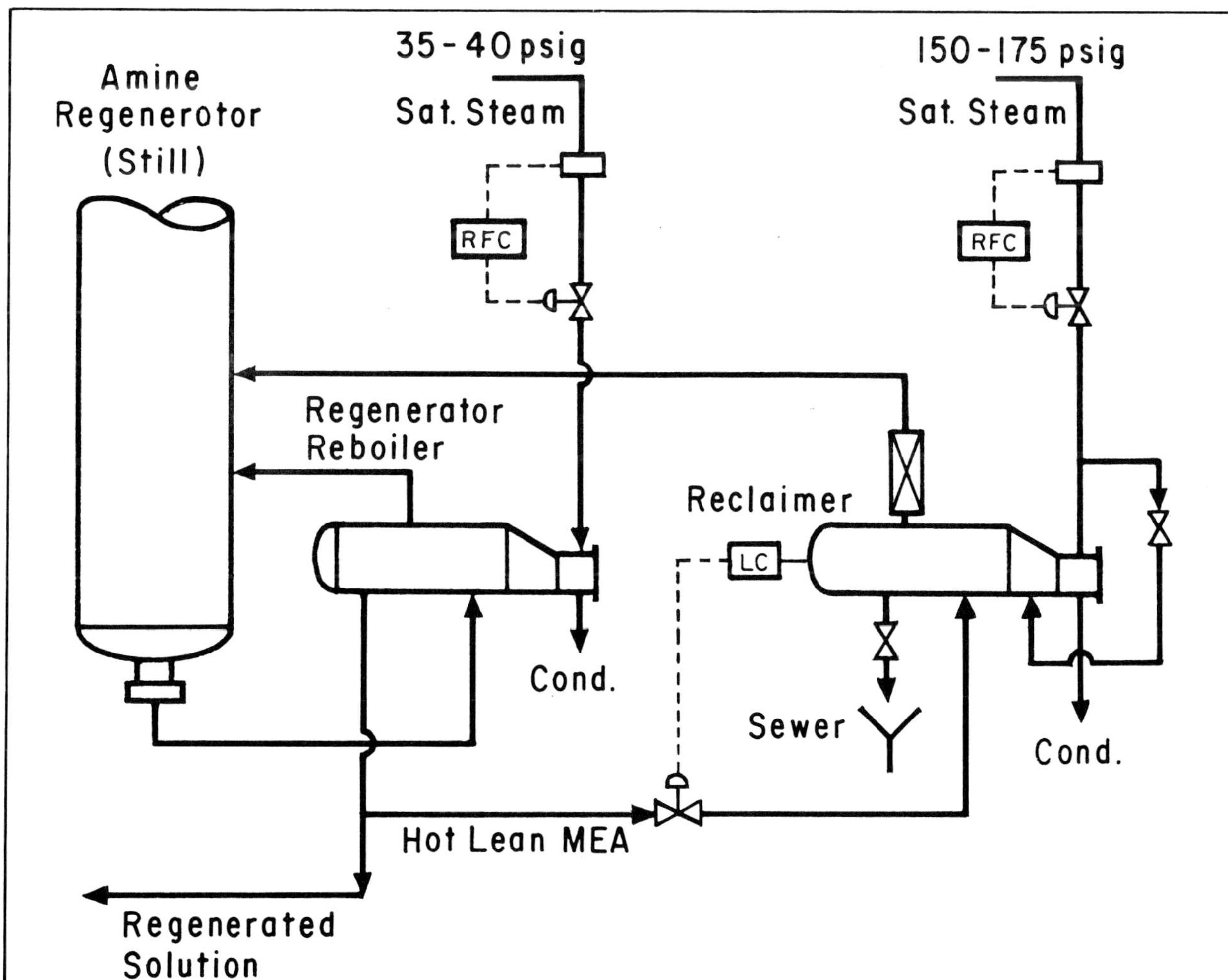

Figure 7.2. Reclaimer—Simplified Flow

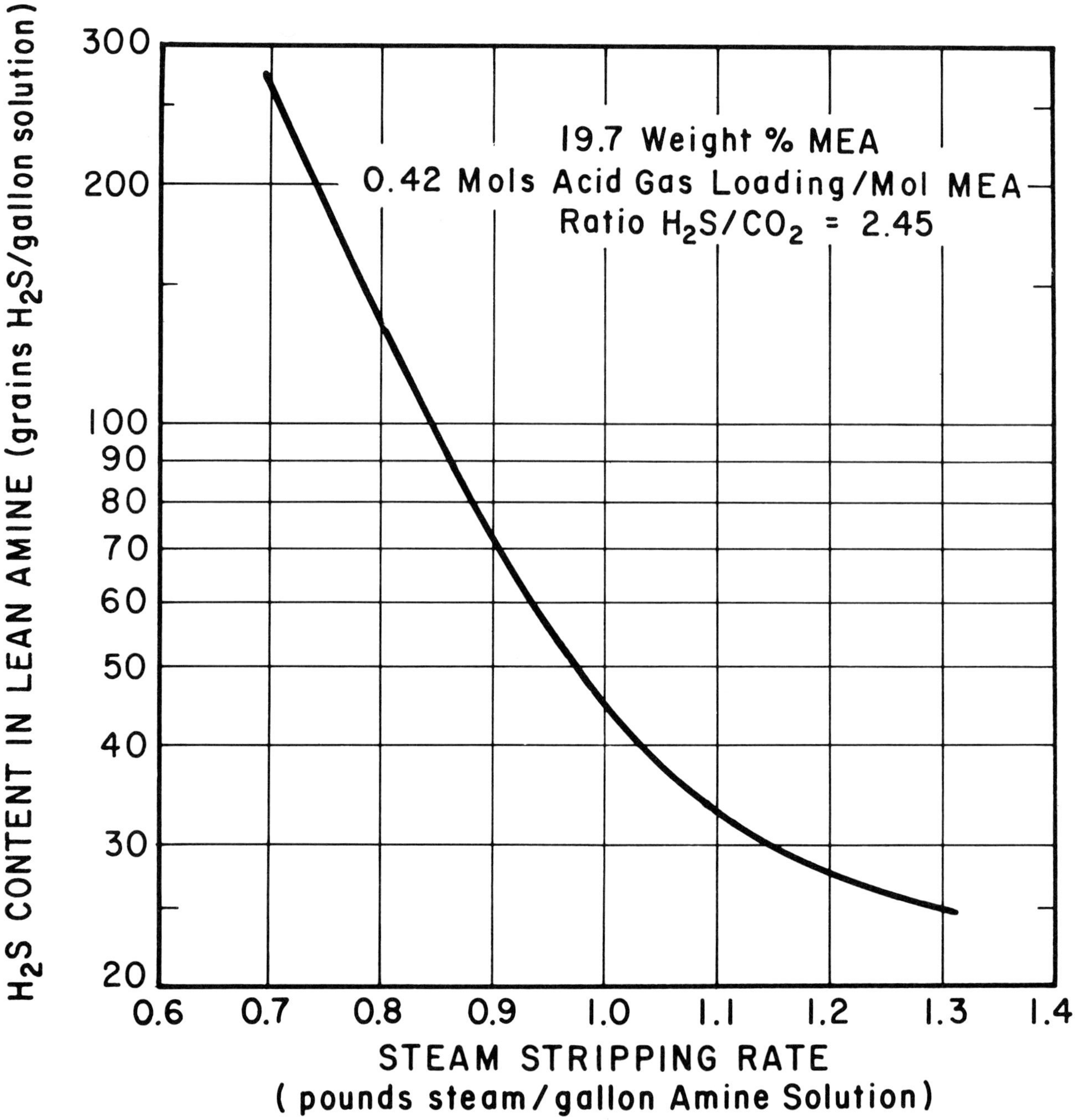

Figure 7.3. Effect of Steam Stripping Rate on Amine Regeneration

aid regeneration, as shown in figure 7.3. If no more heat can be added, try reducing the amine circulation rate. The rate may have been too high for the existing reboiler design. Another consideration concerning reboiler temperature is the difficulty of stripping CO_2 from amine. H_2S is rather easy to strip, but the chemical compound formed when amine reacts with CO_2 must be decomposed before the CO_2 can be stripped. A higher temperature aids the decomposition and may be required even though the higher temperature is detrimental from a corrosion viewpoint.

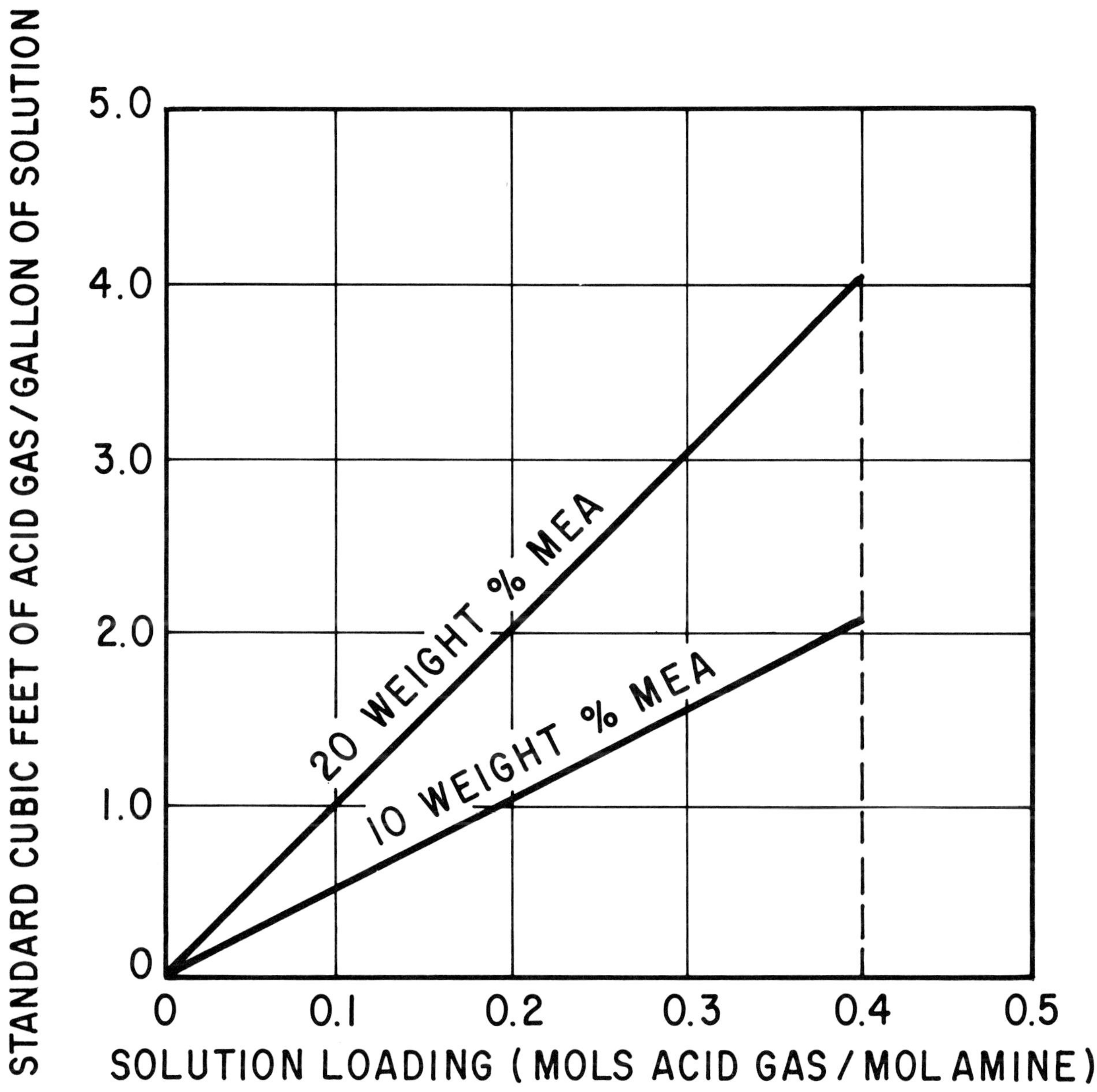

Figure 7.4. Acid-Gas Capacities for MEA Solutions

2. MEA circulation rates and MEA concentrations should be kept within the limits shown in figure 7.4. The MEA concentration should not exceed 20 percent MEA by weight, and the circulation rate should be adjusted to the values shown in figure 7.4. The corresponding loading should not exceed 0.4 mols acid per mol amine (molecular weight of MEA = 61.08).
3. If possible, cool the sour gas to at least 100 F upstream of the contactor. The rich-amine temperature leaving the contactor should not be greater than 120 F. The reason for setting these limits will be seen in the section on corrosion.

Corrosion. Corrosion is a way of life in amine plants, but it can be minimized by the following practices.

1. Keep the amine solution clean. The inlet separator prevents solids from entering the system. Other solids that contribute to corrosion and erosion are removed by the amine filter and reclaimer. The filter should handle up to 10 percent of the circulating system and remove particles down to microns in size. The solids content should be maintained below 0.1 weight percent. The presence of air will cause the amine to degrade into heat-stable salts, so there should be a gas blanket on all amine storage tanks and sumps to exclude air. These heat-stable salts and other MEA degradation products and solids are removed by the reclaimer.
2. Maintain acid-gas loadings within the ranges shown in figure 7.4. Corrosion problems become severe at high temperatures with rich-amine solutions. Most corrosion occurs in areas where the acid gases are actually released from the solution. Severe CO_2 corrosion problems in the reboiler can be minimized by maintaining the solution level above the reboiler tubes.
3. Operate the still at the lowest possible temperatures. Since the reboiler temperature is determined by the boiling point of the amine (see fig. 7.5), this means keeping the amine concentration low and keeping the still pressure low. However, the concentration must be kept high enough to treat the gas, and the still pressure must be kept high enough to maintain flow through the exchangers and filters. The operator should keep the temperature on the heating medium side below 300 F.
4. Maintain a testing program. A regular testing program is important for early detection of problems. Use corrosion inhibitors and run corrosion coupons. Always inspect equipment for corrosion during plant shutdowns. When adding equipment to the system, remember to avoid copper and its alloys. Stress relieve all piping, vessels, and tube bundles.

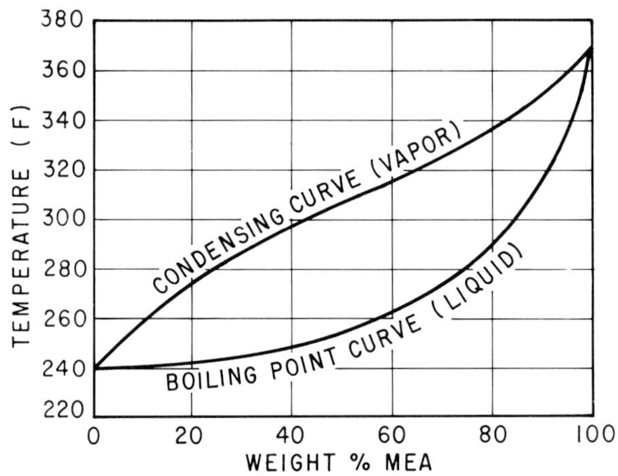

Figure 7.5. Boiling and Condensive Curves for MEA Solutions

Foaming. Foaming is a common problem that will result in a loss of treating capacity and an amine loss. It is usually detected by a sharp rise in pressure drop across the contactor. Foaming is caused by contaminants in the solution and can be prevented in the following ways.

1. Avoid condensation of liquid hydrocarbons in the contactor by keeping the temperature at all spots in the contactor above the hydrocarbon dew point.
2. Keep the contactor clean by water wash or acidizing.
3. Run the reclaimer. Degradation of amine can cause foaming.
4. Keep out field-corrosion inhibitors, soap-base lubricants, and compressor lube oil. Do not overload the inlet separator.
5. Install a charcoal filter to adsorb liquid contaminants.
6. If the preceding remedies fail, then try a defoamer and evaluate the results. Be sure to test the defoamer in a sample bottle before adding it to your plant stream. Keep in mind that just because one brand of defoamer works in one plant does not necessarily mean that it will work in another plant.

Amine consumption. A certain amount of amine will be lost with the residue gas due to vapor pressure of the amine. This should be calculated for each plant and used as a guideline for minimum amine consumption. Figure 7.6 presents the vapor pressure data required to make this calculation. These losses can be kept down by the following processes.

1. Lower the top temperature of the contactor.
2. Water wash the top of the contactor with pure water.
3. Use a good separator with a mist eliminator on the contactor overhead.

The largest amine losses are usually due to occasional carry-over from the contactor due to foaming and continuous small leaks in piping, pump packings, and so forth. One kind of amine loss cannot be prevented. MEA combines with COS and CS_2 to form compounds that cannot be regenerated. These compounds are removed in the reclaimer.

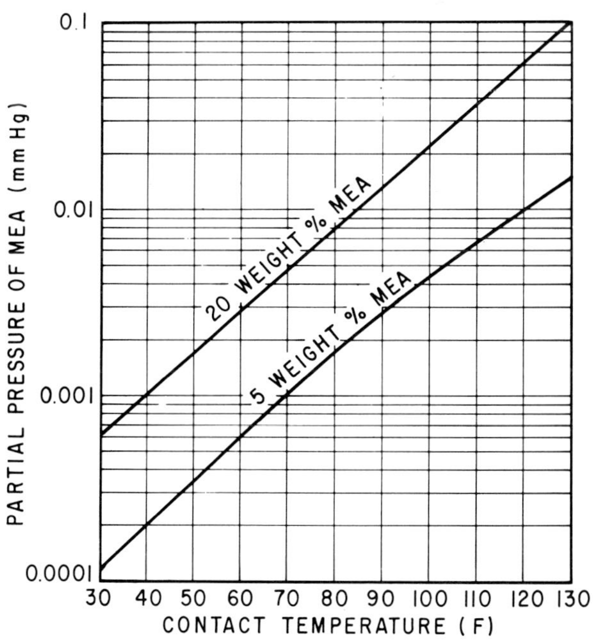

Figure 7.6. MEA Partial Pressure

Winterization. One final operation problem is that an amine system can easily freeze up in winter if the unit is down for long periods of time—see figure 7.7. Thus, the amine system should receive top priority in the event of plant shutdowns during cold weather.

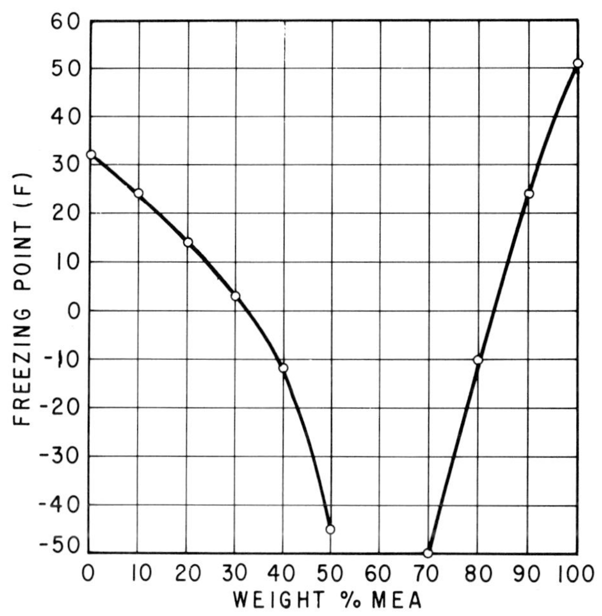

Figure 7.7. Freezing Point Curve for MEA Solutions

Amine Testing

Every plant should set up a routine amine testing program. A number of operating problems can be anticipated and avoided by making the following tests.

1. Observing and smelling the amine reveals problems. A black, green, blue, or amber amine solution indicates corrosion products. Brown indicates that air has entered the system. A strong ammonia odor indicates that the amine has formed degradation products. An oily film indicates hydrocarbon carry-over from the inlet separator or condensation from the inlet gas. If problems are indicated, a complete laboratory analysis should be run.
2. The amine concentration can be measured by the test procedure shown in figure 7.8. Simple laboratory glassware and chemicals are required to run the test.

Test for Free Monoethanolamine In Solution

(Girdler Test SM-1M)

Method

The free monoethanolamine is titrated with 0.1N sulfuric acid using methyl red indicator.

Reagents

0.1N Sulfuric acid

0.1% Methyl red indicator solution

Apparatus

50 ml burette

1 ml Ostwald Folin pipette

250 ml Erlenmeyer flask or beaker

Procedure

Place approximately 100 ml of neutral water in a 250 ml Erlenmeyer flask or beaker. Measure exactly 1 ml of amine solution (note 1) in the Ostwald Folin pipette and transfer the solution into the water by blowing into the pipette. Add 3 drops of indicator solution and titrate to a definite pink endpoint with 0.1N sulfuric acid. Heat to boiling and boil one minute. If sample remains pink, use first endpoint. If it becomes yellow, cool slightly and titrate again to a pink color, taking the final reading for the titration. If in the titration of solutions containing H_2S the color of the indicator is totally destroyed, more indicator should be added.

Calculations

The percent monoethanolamine in solution is calculated from the formula.

$$\text{Grams MEA per 100 ml solution} = 0.611 \times \text{ml } H_2SO_4$$

The percent amine by weight is the value calculated here divided by the specific gravity of the solution analyzed. For plant operation, however, grams/100 ml is satisfactory.

The sample size and the strength and kind of acid used may be varied as desired, provided a corresponding change is made in the factor used in the calculation. The general formula is:

$$\text{Grams MEA per 100 ml solution} = \frac{\text{ml acid} \times \text{N acid} \times 6.11}{\text{volume of sample}}$$

Note 1. With viscous solutions rinse the pipette into the flask with neutral water. Lean (regenerated) amine solution should be titrated wherever possible since the presence of H_2S and CO_2 tends to make the methyl red endpoint somewhat less definite. For control purposes it is ordinarily not necessary to boil the sample as described above after the endpoint has been reached. For accurate results, however, the procedure described above should be followed.

Revised 1-15-46

Figure 7.8. Test for Free MEA in Solution

3. The acid-gas concentration in lean- and rich-amine streams can be determined by the following test procedure.

Test for MEA Acid-Gas Concentration

1. Place 30 to 50 ml of methanol in a flask.
2. Add 1/2 dropper of thymolphthalein indicator, and shake or swirl to mix.
3. Add a 0.5 N caustic solution in just sufficient quantities to obtain a good blue endpoint.
4. Add 1 ml of amine solution, and shake or swirl until the color disappears.
5. Titrate with 0.5 N caustic to the same good blue endpoint as before. Note the volume of caustic required.

$$\text{mol acid gas/mol MEA} = \frac{\text{ml caustic} \times 6.01 \times 0.5(\text{N})}{1\ \text{ml} \times \%\ \text{MEA in solution}}$$

To prepare a 0.5 N caustic solution add 20 grams of caustic pellets to 100 ml distilled water, and stir until dissolved. Add methanol to make total solution of 1,000 ml.

To prepare a thymolphthalein indicator add 1/2 teaspoon thymolphthalein crystals to dropper bottle of methanol.

Safety Considerations

Care should be exercised to prevent contact of the MEA liquid with the eyes. Prolonged or repeated contact of the liquid with the skin should be avoided. In case of contact with the eyes or skin, flush immediately with water. If eye contamination is a possibility, face shields or goggles are suggested. Summon medical aid if the amine enters the eye. At ordinary temperatures, danger from breathing vaporized MEA is slight, but harmful concentrations of the vapor in the air can occur if MEA solutions are heated.

DRY-BED SWEETENING PROCESSES

Dry-bed processes are being successfully used to treat natural gases. The most popular adsorbent is the molecular sieve. Dry-bed sweetening processes are set up in essentially the same way as dry-bed dehydrators (see chapter four). Depending on the type of sieve chosen, the process will remove CO_2, H_2S, and mercaptans selectively. This process is usually highly automated and requires little operator attention. Most operating difficulties center around preserving the useful life of the sieve.

The sieve must be replaced when the treated gas cannot meet minimum sulfur levels within the allowed cycle time. Although sieves lose activity with time, their life is considerably shortened by improper regeneration and contaminants in the gas. To properly regenerate the sieve, the regeneration gas must be at least 550 F, and the bed of absorbent material must be thoroughly heated to an exit temperature of 350 to 400 F. Contaminants such as paraffin, gas corrosion inhibitor chemicals, and oil greatly decrease the life of the sieve. An upstream separator knocks out free liquids and some solids, but a small guard bed of charcoal or activated carbon often gives a surprising increase to sieve life. Normal sieve life should range from 3 to 5 years.

The operator should also periodically get an analysis of the water and CO_2 content of the incoming gas. An increase in either of these components can seriously reduce the ability of the sieve to remove sulfur compounds.

PRODUCT TREATING

Even if the incoming gas is treated, some sulfur compounds will appear in the gasoline and LP-gas products. Product specifications can be met by running the products through a molecular-sieve unit or a caustic unit. For unusual mercaptan problems, further treatment in a Perco process or regenerative caustic process may be required.

By far the simplest unit to install is the solid-bed potassium hydroxide (KOH) treater. This system removes H_2S from both LP-gas and gasoline. All that is required is a single vessel with walnut-sized KOH. Hydrocarbon flow is from bottom to top. No regeneration is required. Water must be added upstream of the unit if the product does not contain moisture already. Approximately 1 qt of water for every 20,000 gal of product is required. The advantage of this system over a caustic (sodium hydroxide) wash is that the liquid product still meets dryness standards. The bed must be drained daily. The proper procedure is to stop all product flow through the bed and allow the system to settle for five minutes. Then begin draining the water, which will take only a few minutes.

An alternate method of H_2S removal is with 13X molecular sieves. This system has an advantage over caustic treating in that sieves can also remove mercaptans and other sulfur compounds, as well as dry the product to meet water specifications. Usually two beds are used; one on adsorption and the other on regeneration. After the adsorption cycle, the bed is drained, depressured, and regenerated with a hot gas stream. These steps give rise to product losses, which is one of the disadvantages of this treating method. The hot bed is usually cooled by injecting liquid product that flashes and carries the heat away.

As with other molecular-sieve processes, the main operating problems center around proper regeneration (inlet temperatures of at least 550 F) and protection from sieve degradation. Normal sieve life is about four years. This can be achieved by placing a guard bed of activated carbon upstream of the unit to remove corrosion inhibitors, amines, absorber oils, glycol, and other sieve contaminants.

The regenerative caustic process is used to remove mercaptans from LP-gas and gasoline, and it has the additional advantage of removing hydrogen sulfide. A 10 percent sodium hydroxide solution is contacted with the liquids in a packed column. The sour caustic is then regenerated in a stripping column. Corrosion is minimized by maintaining the stripper bottoms temperature below 260 F and by blowing down the circulating solution to keep the sodium sulfide content at less than 0.2 percent. Caustic carry-over in the hydrocarbon product is a common problem. It can be removed by installing a sand tower downstream of the caustic unit.

The Perco treater, which utilizes copper chloride, is a more efficient method for removing mercaptans from gasoline. This is another dry-bed process containing copper chloride impregnated on fuller's earth. Hydrogen sulfide, sulfur, and caustic must be removed upstream of the Perco unit because they will deactivate the copper chloride. Free water must also be knocked out since it will wash away the copper chloride. Some water in the gasoline is necessary to maintain the activity. If the gasoline is totally dry, a small sidestream should be run through a pot of water to provide the necessary water. The copper chloride is regenerated by blowing air through the bed. A rule of thumb concerning how much air to use is 1 cu ft of air per barrel of hydrocarbon per 0.01 percent mercaptan sulfur in the hydrocarbon.

ACID-GAS TEST METHODS

The standard method for analyzing the carbon dioxide content of natural gases is with the gas chromatograph (GPA Publication 2261-72). Do not collect the sample by water or glycol displacement since CO_2 is absorbed by these liquids.

Natural gases having an H_2S content greater than 0.5 percent can be collected in a stainless-steel sample bomb and analyzed accurately by a gas chromatograph. However, most natural gases have H_2S contents well below 0.5 percent. Table 4 shows the various measures of H_2S content that are commonly employed; conversions are as indicated.

To obtain an accurate analysis, the gas must be analyzed at the site. The hydrogen sulfide tends to *disappear* into the walls of a sample bomb, whether carbon steel, stainless steel, or glass. This gives a low value for the H_2S content. Table 5 shows suitable test methods for on-site analyses at different levels of H_2S. Note that the cadmium sulfate test also determines the amount of mercaptan sulfur present.

Table 6 lists the standard test methods for sulfur compounds in product streams. There are several words of caution with regard to the copper strip tests. The tests involve exposing a brightly polished copper strip to the liquid product over a certain period of time and temperature. The time and temperature are not the same for the gasoline and LP-gas tests. The product is classified 1, 2, 3, or 4, depending on how tarnished the copper strip becomes. In order to classify the strip properly, it is essential that the tester have a comparison plaque of tarnished copper strips. This is available from ASTM Headquarters, 1916 Race Street, Philadelphia, Pennsylvania 19103. Finally, the most critical part of the test is the surface preparation and polishing of the strip. This requires extreme care and patience to produce a good test.

There are no corresponding industry standards for carbon dioxide in products. Normally, CO_2 is a major concern with ethane products. The usual test method is with a gas chromatograph either on a liquid sample or a gas sample. The sample should *not* be collected by water displacement or glycol displacement since carbon dioxide is appreciably absorbed by the water or glycol. Mercury displacement methods may be used. Also, the newer piston-type sample bombs or bladder bombs provide an adequate sample.

Table 4. Conversion Factors for Various Measures of H_2S Content in Natural Gas

H_2S: 1.0%	= 628 grains per 100 scf
1.0%	= 10,000 ppm
0.25 grains per 100 scf	= 4.0 ppm by volume
0.25 grains per 100 scf	= 0.0004% by volume

Table 5. H_2S Test Methods for Gases

Method	*Reference*	*H_2S Range (grains/100 scf)*
Lead Acetate (Qualitative)	GPA Plant Operations Manual—Test B-14	Presence of more than 0.05 can be indicated
Cadmium Sulfate—Iodometric Titration	GPA 2265-68	0-5
Colorimetric Sniffer Tube	Mine Safety Appliance Company	0-400
Methylene Blue	IGT Research Bulletin 5	0-1000

Table 6. Product Test Methods

	Propane, Butane	*Gasoline*
H_2S Corrosion	Copper Strip Method (GPA 2140-73)(ASTM D1838-64)	Copper Strip Test (ASTM D130-68)
Total Sulfur	Oxy-hydrogen Burner (GPA 2140-73) (ASTM D2784-70)	
Mercaptans		Doctor Test (GPA 1138)

Chapter VIII

Sulfur Recovery

During the early phases of the natural gas industry, only sweet gas—that is gas free of CO_2, H_2S, and other sulfur compounds—was used as fuel. As the demand for gas grew, it became necessary to turn to sour gas for more supply and, hence, to provide means by which these impurities could be removed. Processes to remove these impurities were developed and utilized by the petroleum industry. The impurities were burned or vented to the atmosphere. The next step was to employ a process by which H_2S and other gaseous sulfur compounds could be converted to elemental sulfur. C. F. Claus, an English scientist, had discovered and patented in 1883 the reaction between H_2S and sulfur dioxide (SO_2), which produces elemental sulfur. Today, the Claus process is the most widely used as well as the most reliable method for the conversion of sulfur compounds in acid gas to elemental sulfur.

For many years, small sulfur plants, which could produce 10 long tons per day or less, were not built because of the relatively high investment cost and the low cash return that could be expected. During the mid and late 1960s, the demand for sulfur throughout the world increased dramatically. The price increases that accompanied these demands spurred the construction of sulfur plants of all sizes throughout North America. Sulfur prices have since declined sharply, but sulfur plants continue to be built as the demand for natural gas increases. State and federal clean air laws have resulted in either the construction of new sulfur plants or the modification of existing ones to comply with strict emission standards.

THE CLAUS PROCESS

The Claus process is accomplished by supplying the correct amount of oxygen to oxidize the hydrogen sulfide to sulfur and also to burn the hydrocarbon present in the acid-gas feed to the sulfur plant. Reactions take place in the sulfur furnace as well as in the first, second, and, in some cases, the third catalytic converter. The most common catalysts are activated natural bauxite or manufactured alumina. For these reactions to take place under conditions that will produce an optimum recovery, it is necessary to control closely the reaction temperature and air rates to the burner. The reactions that occur in the furnace or fire tube are—

$$H_2S + O = S + H_2O$$

$$2\ H_2S + 3\ O_2 = 2\ SO_2 + 2\ H_2O$$

$$2\ H_2S + SO_2 = 3\ S + 2\ H_2O$$

The reaction that occurs in the furnace and catalyst beds is—

$$SO_2 + 2\ H_2S = 3\ S + 2\ H_2O$$

Up to 70 percent of the sulfur is formed in the furnace, and the remainder is formed in the catalyst beds. A Claus plant of proper design and three reactors can achieve up to a 98 percent yield of sulfur, but, in most two-reactor plants, approximately a 94 percent to 96 percent conversion for an acid gas that is mostly hydrogen sulfide will be obtained. A photograph of two sulfur catalytic reactors is shown in figure 8.1.

SULFUR PLANT PROCESS FLOW

Acid-gas composition and volume vary so widely from field to field that sulfur plants, for the most part, are designed for a specific location. Sulfur plants operate at very low pressures. Pressures normally vary between 2 psig and 10 psig depending upon size and design. The temperatures vary slightly with acid-gas composition and plant design. Figure 8.2 illustrates the equipment arrangement in a package-type sulfur plant. Figure 8.3 is a flow diagram for an articulated sulfur plant.

The following description of a Claus sulfur plant is generally typical, but there can be considerable variation in operating temperatures. The acid gas flows into a specially designed burner located at the front end of a refractory-lined furnace. Air is admitted in carefully controlled quantities to produce the most sulfur from the total hydrogen sulfide. Too much or too little air reduces the yield. This combustion produces temperatures ranging from 1,900 F to 2,400 F. The hot gas flows from the furnace into

Figure 8.1. Sulfur Catalytic Reactors

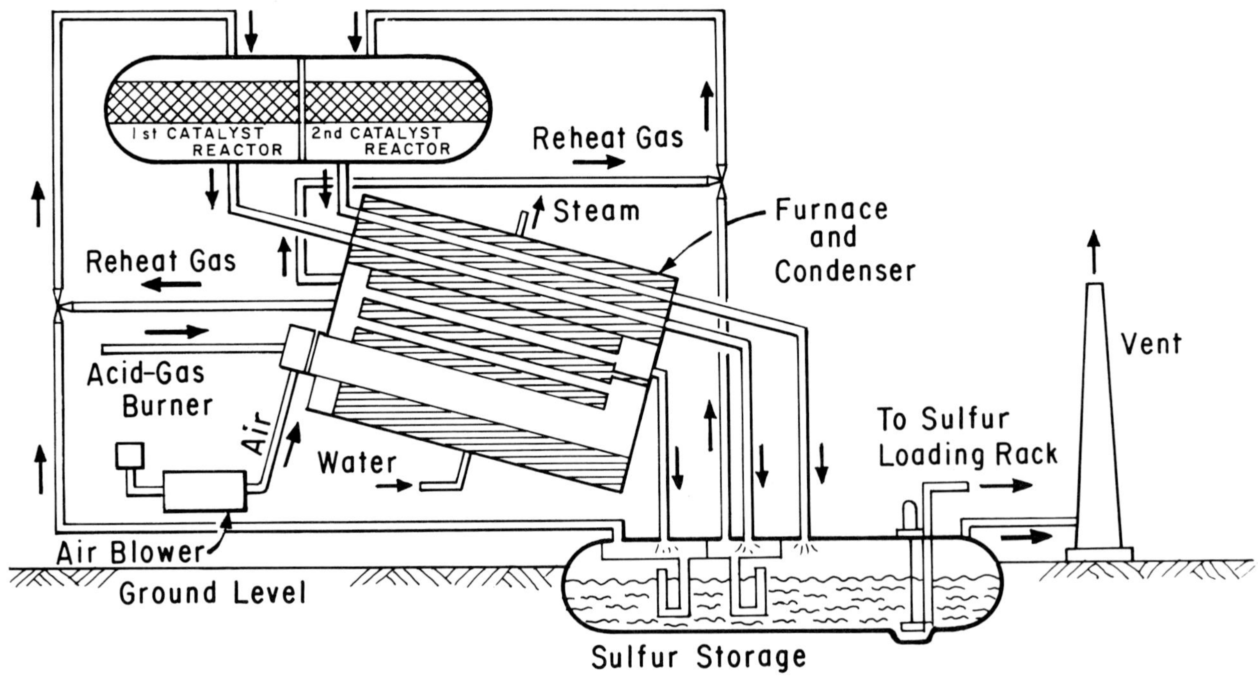

Figure 8.2. Package-Type Sulfur Plant Equipment

a boiler, commonly referred to as a waste heat boiler. The gas is cooled to approximately 900 F in the first pass and approximately 650 F in the second pass. It flows from the second pass through a sulfur condenser where elemental sulfur in the vapor phase, formed in the furnace or fire tube in the boiler, is condensed to liquid sulfur. The liquid sulfur is withdrawn at this point, and the gas containing unreacted sulfur compounds flows to the first catalytic reactor. The temperature of the gas leaving the first condenser is approximately 325 F. This gas must be heated to approximately 440 F prior to entering the first catalytic reactor if the proper reaction is to take place. The reheating of this gas is accomplished by bypassing a portion of the hot (900 F) gas from the first pass of the waste heat boiler. The gas enters the first catalytic reactor at approximately 440 F and leaves at approximately 580 F. It then flows through a condenser where liquid sulfur is again separated from the gas containing unreacted H_2S and SO_2.

It is necessary to reheat the gas before it enters the second catalytic reactor. The reheating can be accomplished by either bypassing hot gas from the first pass of the boiler or by a reheat exchanger utilizing hot gas from the first reactor outlet prior to its entry into the condenser. The gas enters the second catalytic reactor at approximately 420 F and leaves at approximately 470 F. After passing through another condenser where it is cooled and again the liquid sulfur is separated, the unreacted gas can pass either to a third catalytic reactor to increase recovery or go directly to a vent stack or an incinerator.

The conversion of hydrogen sulfide to elemental sulfur is an *exothermic reaction,* which means it produces heat. It will be noted that each successive catalyst bed has a smaller temperature rise across it (outlet temperature minus inlet temperature). This indicates that the first bed is converting more hydrogen sulfide and sulfur dioxide to elemental sulfur than is the second bed. If the temperature rise of the first bed decreases and the second bed increases, activity in the first catalyst bed has decreased. This condition can be caused by the inlet temperature to the catalyst bed being too low, resulting in sulfur condensation on the beds and a decrease in sulfur recovery. If this condition occurs, it will be necessary to heat the bed to approximately 500 F and hold it at this temperature for 24 hours to remove the sulfur from

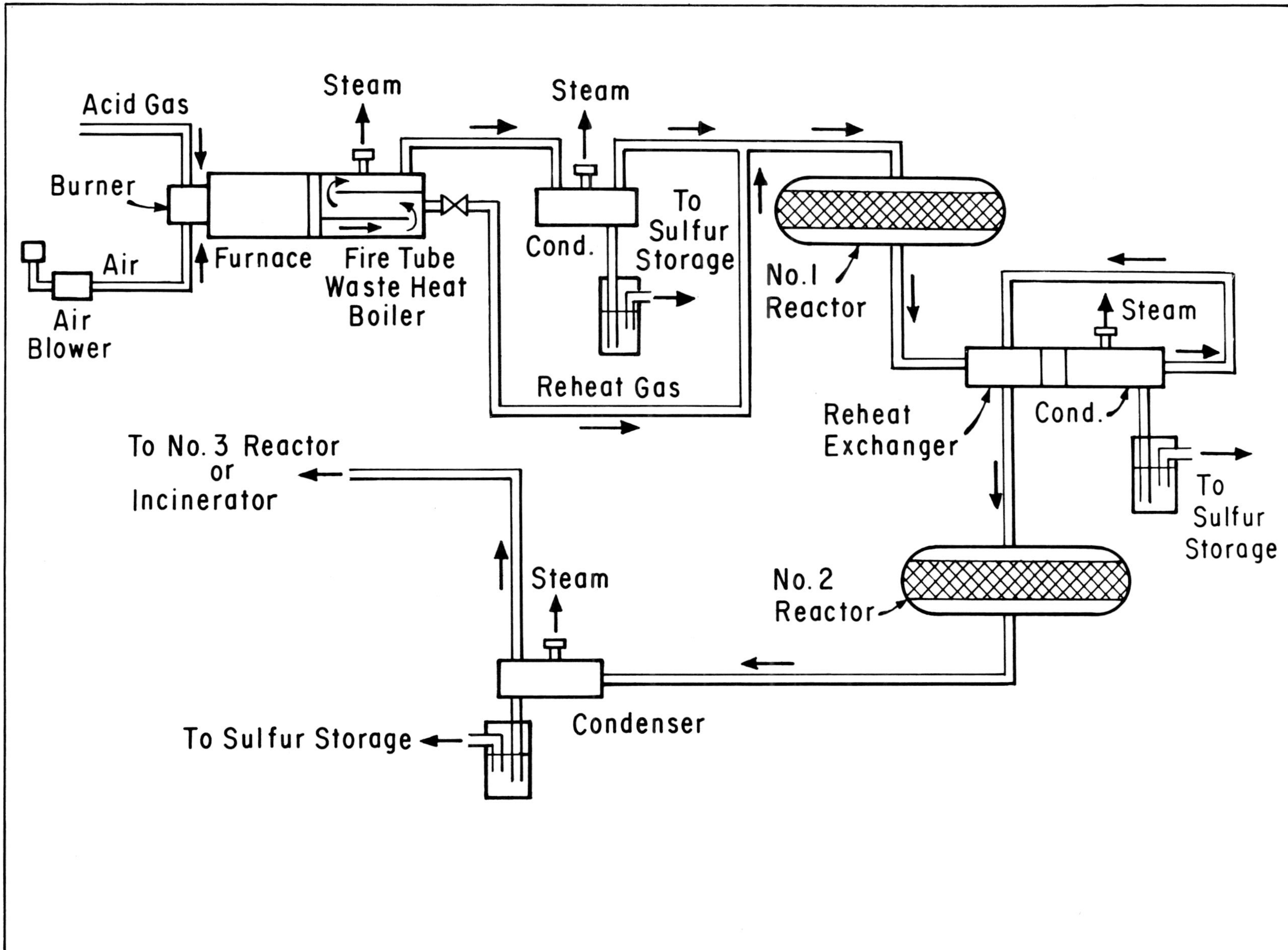

Figure 8.3. Flow Diagram of an Articulated Sulfur Plant

the beds. This usually restores the bed's activity and removes the condensed sulfur.

Sulfur melts at approximately 240 F, and sulfur produced by the Claus process is in the liquid phase. Claus sulfur plants are designed so that all liquid sulfur drains from the plant to a heated storage tank, which is usually below ground level. Liquid sulfur from the storage tank can be pumped to above-ground liquid storage, to solid storage blocks, to tank trucks, or tank cars.

SULFUR PLANT START-UP

The sulfur furnace, the reactors, and all process streams must be brought up to operating temperature prior to admitting hydrogen sulfide to the burner. The sulfur plant is preheated by burning fuel gas in the sulfur furnace burner. To avoid damage to the burner or refractory from high combustion temperature, the fuel should be diluted with some inert gas such as nitrogen or carbon dioxide. The fuel gas-air ratio must be closely controlled because excess air causes sulfur fires, and deficient air results in incomplete combustion that will cause carbon deposition and thereby reduce catalyst activity. When the temperatures are at the normal operating level, the sulfur plant is ready to receive hydrogen sulfide. Acid gas is introduced in small increments, and the fuel gas is simultaneously cut back by a like amount. When all fuel gas has been cut out, the air is adjusted to maintain the correct H_2S - SO_2 ratio, and hot-gas bypasses are adjusted to maintain the correct reactor inlet temperatures. When these steps are taken, the sulfur plant should be operating smoothly and efficiently.

PLANNED SHUTDOWN FOR EXTENSIVE REPAIRS

Since the pipes, reactors, and condensers contain sulfur, it is very important to remove this sulfur before allowing the sulfur plant to cool. If the sulfur is not removed, it will condense, and it could plug lines and catalyst beds, resulting in expensive repair work. Therefore, in a planned shutdown, the hydrogen sulfide flow to the burner should be reduced and the air increased until nearly all of the hydrogen sulfide is burned to sulfur dioxide. It is important to avoid excess air, which causes sulfur trioxide (SO_3) to form. SO_3 in turn causes severe corrosion. By converting almost all the hydrogen sulfide to sulfur dioxide, little reaction will occur to produce additional sulfur vapors. The sulfur dioxide sweeps the majority of the sulfur vapors from the sulfur plant process streams. It should be noted that acid gas composed largely of H_2S can burn as hot as fuel gas and damage refractory unless very low flow rates are used. High temperatures can be overcome by utilizing an inert gas such as nitrogen or carbon dioxide to quench the flame temperature. It may be advantageous in some cases to purge the entire sulfur plant after the all-SO_2 cycle with an inert gas to avoid SO_2 corrosion.

SULFUR PLANT CONTROLS

The instrumentation for a sulfur plant is primarily concerned with the close control of the air-gas ratio. This control is supplied by a ratio controller that measures the acid-gas flow to the burner and automatically corrects the air flow as the acid-gas flow changes. However, if the hydrogen sulfide or water content of the acid gas changes, a different ratio would be required. To determine when this is necessary, the tail gases from the sulfur plant should be analyzed to determine the H_2S - SO_2 ratio. The ratio should be maintained at 2 to 1. If the hydrogen-sulfide-to-sulfur-dioxide ratio is less than 2.0, then excess air is being used, and the air flow must be reduced. Conversely, when the hydrogen-sulfide-to-sulfur-dioxide ratio is greater than 2.0, there is an air deficiency. Analyses should be run frequently to ensure plant efficiency. Adjustments to the ratio controller can be made after each test to maintain the correct ratio. The method of tail-gas analysis most widely used is the Tutwiler analysis using a special burette with iodine and caustic solutions. With more and more emphasis being placed on the abatement of air pollutants, the use of automatic analyzers for tail-gas analyses is becoming more common. A chromatograph or ultraviolet analyzer for H_2S and SO_2 can be used to analyze the tail gas and automatically reset the ratio controller, or it can be used to indicate the ratio, and the operator can make adjustments to the ratio controller.

TAIL-GAS CLEANUP

With the advent of more stringent air pollution laws, demands are being made for higher performance sulfur plants. These demands are being met by utilizing what is referred to as tail-gas cleanup. The tail gas from a Claus plant is processed through a separate process to remove the majority of the remaining sulfur compounds. Several of these processes are commercially proven, and others are under development, and still others are on the drawing board. The Wellman-Lord, I.F.P., Beavon, and Sulfreen processes are operating, and the SCOT and Cleanair are starting commercial construction.

References for Chapter 7

1. Allied Chemical Corporation. "The Selexol Process." Bulletin. Morristown, New Jersey.
2. Allied Chemical Corporation: Solvay Process Division. *The Hot Potassium Carbonate Process for Acid Gas Absorption.* Technical Service Report 6.61. Syracuse, New York, 1961.
3. Ballard, D. "How to Operate an Amine Plant." *Hydrocarbon Processing,* April, 1966, p. 137.
4. Buckingham, Paul A. "Fluor Solvent Process Plants: How They Are Working." *Hydrocarbon Processing,* April, 1964, p. 113.
5. Burnham, J. G. "Ten Methods of Sour Gas Processing." *Oilweek,* May 1, 1961, p. 38.
6. Campbell, John M. *Gas Conditioning and Processing.* Norman, Oklahoma, 1970.
7. Conviser, S. A. "Molecular Sieves Used to Remove Mercaptans from Natural Gas." *Oil and Gas Journal,* December 6, 1965, p. 130.
8. Dailey, L. W. "Present Status of the SNPA-DEA Process." *NGPA Annual Convention Proceedings, 1970.*
9. Dingman, J. C. and Moore, T. F. "Compare DGA and MEA Sweetening Methods." *Hydrocarbon Processing,* July, 1968, p. 138.
10. Dingman, J. C. et al. "Minimize Corrosion in MEA Units." *Hydrocarbon Processing,* September, 1966, p. 285.
11. Dow Chemical Company. *Gas Conditioning Fact Book.* Midland, Michigan, 1962.
12. Dunn, C. L. et al. "First Plant Data from Sulfinol Process." *Hydrocarbon Processing,* April, 1965.
13. Eckert, J. S. "Sour Gas Amine Scrubbers and Regenerators Using Packed Towers." *Canadian Gas Journal,* March-April, 1972, p. 36.
14. Eickmeyer, A. G. "Acid-gas Removal." *The Oil and Gas Journal,* September 22, 1958, p. 106.
15. "Estasolvan: New Gas Treating Process." *Hydrocarbon Processing,* May, 1970, p. 145.
16. Fitzgerald, K. J. and Richardson, J. A. "How Gas Compostion Affects Treating Process Selection." *Hydrocarbon Processing,* July, 1966, p. 125.
17. Gustafson, K. J. and Healey, M. J. "Removal of Hydrogen Sulfide from Natural Gas/Carbon Dioxide Mixtures with Molecular Sieves." Gas Conditioning Conference. Norman, Oklahoma, April, 1968.
18. Goar, B. G. "Today's Gas Treating Processes—Part I." *The Oil and Gas Journal,* July 12, 1971, p. 75.
19. Goar, B. G. "Today's Gas Treating Processes—Part II." *The Oil and Gas Journal,* July 12, 1971, p. 84.

20. Hall, G. D. and Polderman, L. D. "Design and Operating Tips for Ethanolamine Gas Scrubbing Systems." *Chemical Engineering Progress,* October, 1960.
21. Harris, T. B. "Natural Gas Treating with Molecular Sieves." *Pipeline and Gas Journal,* August, 1972, p. 40.
22. Hegwer, A. M. and Harris, R. A. "Selexol Solves High H_2S/CO_2 Problem." *Hydrocarbon Processing,* April, 1970.
23. "Jefferson Gas Treating Briefs." Jefferson Chemical Company. Houston, Texas, 1964.
24. Lurgi Gesellschaft Für Wärme-Und Chemotechnik Gmbh. "Purisol for Gas Treating." Bulletin.
25. Miller, S. G. and Robuck, R. D. "The Stretford Process at East Wilmington Field." *Journal of Petroleum Technology,* May, 1972, p. 545.
26. Maddox, R. N. "Conventional Sweetening Processes." *NGPA Annual Convention Proceedings,* 1970, p. 98.
27. National Tank Company. "Gas Treating and Sulfur Recovery." CE Natco Bulletin. Tulsa, Oklahoma, 1971.
28. *NGPSA Engineering Data Book.* 1972.
29. Smith, R. F. "Tips on DEA Treating." *Hydrocarbon Processing,* July, 1972, p. 98.
30. Swaim, Jr., C. D. "Gas Sweetening Processes of the 1960's." *Hydrocarbon Processing,* March, 1970.
31. Union Carbide Corporation. "Gas Treating Chemicals." Bulletin. New York, 1969.
32. Wendt, Jr., C. J. and Dailey, L. W. "Gas Treating: The SNPA Process." *Hydrocarbon Processing,* October, 1967, p. 155.